TOP GUNS
& TOXIC WHALES
The Environment & Global Security
Gwyn Prins & Robbie Stamp

EARTHSCAN PUBLICATIONS LIMITED, LONDON

TOP GUNS
& TOXIC WHALES
The Environment & Global Security
Gwyn Prins & Robbie Stamp

For Elizabeth, Samuel and Susannah

First published 1991
Earthscan Publications Ltd
3 Endsleigh Street, London WC1H 0DD

Copyright © 1991 Gwyn Prins and Robbie Stamp
Central Logo Copyright © 1989 Central Independent Television plc

All rights reserved

British Library Cataloguing in Publication Data
Prins, Gwyn
   Top Guns and Toxic Whales:
   The Environment and Global Security
   1. Environment. Degradation
   I. Title II. Stamp, Robbie
   333.7

ISBN 1-85383-094-1

This publication accompanies the television documentary *Top Guns and Toxic Whales*, produced by Central Independent Television in association with the Better World Society and the Television Trust for the Environment as part of Central ITV's Viewpoint '91 series.

Production by Bob Towell
Typeset by Neat Graphics
Manufactured in Italy by Grafedit S.p.A., Bergamo

Earthscan Publications Ltd is an editorially independent subsidiary of the International Institute for Environment and Development (Charity No. 800066).

Photographs and illustrations used with the kind permission of the following:
Chapter 1: Greenpeace; Central Independent Television plc; United States Navy; Dr A. R. Martin, Sea Mammal Research Unit; Rijksmuseum, Amsterdam; Edward S. Curtis, Portraits from North American Life; David King
Chapter 2: Mary Evans Picture Library; North American Aerospace Defense Command; Lawrence Moore, © Central Television; Missouri Department of Revenue; Elizabeth L. Venrick, Scripps Institute of Oceanography; Charles Secrett/Friends of the Earth
Chapter 3: Lawrence Moore, © Central Television; National Maritime Museum; Imperial War Museum; Don McCullin
Chapter 4: Bruno Sorrentino, © Central Television; Lawrence Moore, © Central Television; CIA
Chapter 5: National Snow and Ice Data Center (NSIDC), University of Colorado, Boulder, CO; International Atomic Energy Agency; Bill Pierce, Colorific; Edinburgh University Wavepower Project; Philips
Chapter 6: Brian Moser
Chapter 7: Greenpeace; United States Navy; Institute of Oceanographic Sciences, Deacon Library; Commander in Chief, Royal Navy; Rolls Royce plc; Dr A. R. Martin, Sea Mammal Research Unit

Previous page: **The F14A Tomcat is one of the United States Air Force's most sophisticated planes, flown by an elite group of pilots known as "Top Guns".** *Below:* **Beluga whales.**

Contents

Acknowledgements 6
Metric/Imperial Conversion Tables 8
Glossary of Acronyms 9
Introduction 11

CHAPTER 1:
You Can't Shoot an Ozone Hole 12
   The Tomcat and the pram 12
   Breakdown of the social contract 16
   Redefining security 21
   Leviathan meets Humpty Dumpty 29

CHAPTER 2:
Visiting CASSANDRA 34
   The War Room 34
   The screens of CASSANDRA 36
      Human population growth 40
      Man-made poisons 44
      The hole in the sky 46

CHAPTER 3
"De Nile Ain't Jus' a River in Egypt . . ." 52
   Four reasons for disbelieving
      CASSANDRA 52
   The perils of De Nile 61
   The shadow of the "grey zone" 62
   Global Poohsticks 68
   Observe! Analyse! Act! 69

CHAPTER 4
The Worst of Times 76
   Revolution's choices 76
   Oil wars 80
   The invisible worm that flies in
     the night 84
     A lost country 85
   Run For Your Life! 90

   "A well founded fear" 92
   Fatima's march 93
   To reach El Norte! 93
   Water wars 97
     Operation PHARAOH 100

CHAPTER 5
The Best of Times 106
   Were the doomsters wrong? 106
   Energy is our guide 108
   Towards a global benchmark
     for sustainability 112
   Four stages in energy history 113
   Enter the negawatt 116
   Developing countries: leapfrogging
     into the future 125

CHAPTER 6
So What's Holding Us Back? 132
   Poor people's chains 136
   The chains of the rich 139

CHAPTER 7
Captain Cook to the Rescue 142
   A transitional world 142
   Immediate needs; immediate threats 146
   Making spinoffs spin 151
     BIG/ISTIG in the North 152
     BIG/ISTIG in the South 153
   A cautionary tale before we get too
     excited 156
   Making it safe to cash in the peace
     dividend 160

Conclusion 162
Further reading 165

Acknowledgements

We wish to express our gratitude to the many organizations and individuals who have helped to make this book possible. To Central Television and Television for the Environment, whose long-term commitment has underlain the project of which this book is a part, and the Better World Society, in partnership with the Carter Center, the American co-producers of the documentary film *Top Guns and Toxic Whales*, which is the companion of this book, was the stimulus for writing it and upon which it draws for much case material. We are grateful to all the people who were interviewed for the film, many of whom appear in the book.

At Central special thanks go to Roger James and Richard Creasey, at TVE to Robert Lamb. We are particularly grateful to Julie Stoner, the film's Production Assistant. Kim Johnson and Thomas Harding helped with research. Tom Belford and Vicky Markell at the Better World Society were splendid partners. Chris Holifield was an exemplary agent. The staff of Earthscan, with David King and Judy Groves, turned the manuscript into a book in record time.

In its style and treatment, the film *Top Guns and Toxic Whales* exemplifies a new approach to making TV documentaries about the environment. It is the third in a trilogy of which the first, which won a Prix Italia in 1990, was called *Can Polar Bears Tread Water?* It dealt with the subject of global warming. The second, entitled *When the Bough Breaks*, looked at the plight of children in degraded environments today and at their peculiar vulnerability to worsening conditions. Lawrence Moore has been Producer/Director of the trilogy, and their innovative style and treatment reflect his vision. We are especially grateful to him for his unstinting assistance with the graphics and illustrations for this book, for his encouragement and support.

However, in terms of the book, there would have been no physical manuscript for the Publisher's deadline had it not been for one person. Dee Noyes, Gwyn Prins's Assistant and Programme Secretary of the University of Cambridge Global Security Programme dropped everything and typed the bulk of the text in ten days. Our debt to her is enormous; and our thanks also to her husband, Andrew, for technical assistance at a critical moment! The Global Security Programme is the University's spearhead for teaching and research on the new security agenda of which environmental security is a part. It is newly founded, and its initial phase is funded largely by the John D. and Catherine T. MacArthur Foundation of Chicago. Without the Foundation's generosity, neither Dee nor the Programme would be there; so we wish to acknowledge our gratitude to it.

Parts of this book draw upon ideas which Gwyn Prins has developed over recent years for the British Armed Forces Staff Colleges. In 1988-89, he assisted the Army in redesigning the extra-European and environmental security portion of the Army Staff Course at Camberley; ideas which were also explored up the road at the

Acknowledgements

RAF Staff College, Bracknell, over several years at the Joint Services Defence College at Greenwich and at the Universitat der Bundeswehr at Neubiberg, Munich. He is grateful particularly to the Higher Command and Staff Courses, 1989 and 1990 for trenchant criticism, to the then Commandant of the Staff College, Major General Jeremy MacKenzie and the then Director of the HCSC, Major General Rupert Smith.

However, his longest association in these matters has been with the Royal Navy. He wishes to express particular thanks to successive Royal Naval Staff Courses and Initial Staff Courses at the Royal Naval Staff College, Greenwich, for their stimulus in questioning; to Geoffrey Till and the academic staff at RNSC and above all, in his former post as the Staff College's innovative Commandant during the crucial years of the mid 1980s, to Commodore Jeremy Blackham

When these matters moved into the policy discussion arena during 1990, the fraternal "Cambridge squared" link between the Global Security Programme in Britain and the Committee on International Security Studies at the American Academy for Arts and Sciences in Cambridge, Massachusetts operated profitably, ensuring a proper Anglo-American discussion, thanks to the help of Dr Jeffrey Boutwell, CISS Director. In Cambridge (England), Gwyn Prins is indebted to Professor Martin Rees of the Institute of Astronomy for help in arranging a Saltmarsh Science Policy Meeting, to the Director of the Marine and Atmospheric Science Directorate at the Natural Environment Research Council, Dr John Woods, and to the Chief Scientific Adviser at the Ministry of Defence, Professor Ron Oxburgh. Gwyn Prins has profited from many enlightening discussions with Bill Hopkinson, Head of the Arms Control and Disarmament Unit at MoD. The sympathetic interest of the First Sea Lord, Sir Julian Oswald, was greatly appreciated, as was the continuing support of Commodore Blackham, now serving as the Royal Navy's Director of Plans and Programmes. The Service is fortunate to have officers of such calibre in senior command at this time.

In preparing the book, another constituency was also of great assistance. David Gee, Director of Friends of the Earth UK, has been liberal in his assistance both at a planning meeting for the television part of the project, held at Minster Lovell Mill in the spring of 1990, and later in many ways. Jeremy Leggett, Director of Oceanographic and Atmospheric Science at Greenpeace International, provided an important and thought-provoking paper at Minster Lovell, and has been generous in affording us access to scientific reference and papers, not least in the authoritative collection on Global Warming which he edited. We are grateful to all the colleagues who attended that weekend in Oxfordshire, one of whom, Hal Feiveson, was a representative of the Centre for Energy and Environment Studies at Princeton and upon whose work this study has drawn extensively. Needless to say, responsibility for the analysis made here does not lie with any of these many colleagues and institutions.

While, as always in Gwyn Prins's dealings with the Forces over many years, much of what is written here will doubtless shock the conventional military mind (as indeed seen from another direction, it may surprise many environmentalists) it is hoped that both groups may derive more profit from what follows than be discomforted by the experience!

Finally, we wish to thank our families for putting up with the vast intrusions upon the Christmas holiday which the writing of this book entailed. It is dedicated to our children.

Metric/Imperial Conversion Tables

Weights and Measures

Length
1 millimetre = 1000 micrometres = 0.0394 inch
1 centimetre = 10 millimetres = 0.3937 inch
1 metre = 100 centimetres = 1.0936 yards
1 kilometre = 1000 metres = 0.6214 mile
1 inch = 2.54 centimetres
1 foot = 12 inches = 30.48 centimetres
1 yard = 36 inches = 0.9144 metre
1 mile = 1760 yards = 1.6093 kilometres

Area
1 sq metre = 10,000 sq centimetres = 1.1960 sq yards
1 hectare = 10,000 sq metres = 2.4711 acres
1 sq kilometre = 100 hectares = 0.3861 sq mile

1 sq foot = 144 sq inches = 0.0929 sq metre
1 sq yard = 9 sq feet = 0.8361 sq metre
1 acre = 4840 sq yards = 4046.9 sq metres

Capacity
1 cu decimetre = 1000 cu centimetres = 0.0353 cu foot
1 cu metre = 1000 cu decimetres = 1.3080 cu yards
1 litre = 1 cu decimetre = 0.22 gallon

1 cu yard = 27 cu feet = 0.7646 cu metre
1 pint = 4 gills = 0.5683 litre
1 gallon = 8 pints = 4.5461 litres

Weight
1 gram = 1000 milligrams = 0.0353 ounce
1 kilogram = 1000 grams = 2.2046 pounds
1 tonne = 1000 kilograms = 0.9842 ton
1 ounce = 437.5 grams = 28.35 grams
1 pound = 16 ounces = 0.4536 kilogram
1 stone = 14 pounds = 6.35 kilograms
1 ton = 2240 pounds = 1.016 tonnes

USA Dry Measure Equivalents
1 pint = 0.9689 UK pint = 0.5506 litre
1 bushel = 0.9689 UK bushel = 35.2380 litres

USA Liquid Measure Equivalents
1 pint (16 fl oz) = 0.8327 pint = 0.4732 litre
1 gallon = 0.8327 gallon = 3.7853 litres

Temperature Conversion
$C = 5/9 (F - 32)$ $F = 9/5 C + 32$

Celsius   −18° −10 0   10   20   30   40

Fahrenheit   0° 10 20 32 40 50 60 70 80 90 100

Conversion Formulae

To Convert	Multiply by
Inches to Centimetres	2.540
Centimetres to Inches	0.39370
Feet to Metres	0.3048
Metres to feet	3.2808
Yards to Metres	0.9144
Metres to Yards	1.09361
Miles to Kilometres	1.60934
Kilometres to Miles	0.621371
Sq Inches to Sq Centimetres	6.4516
Sq Centimetres to Sq Inches	0.15499
Sq Metres to Sq Feet	10.7638
Sq Feet to Sq Metres	0.092903
Sq Yards to Sq Metres	0.83612
Sq Metres to Sq Yards	1.19599
Sq Miles to Sq Kilometres	2.5899
Sq Kilometres to Sq Miles	0.386103
Acres to Hectares	0.40468
Hectares to Acres	2.47105
Cu Inches to Cu Centimetres	16.3870
Cu Centimetres to Cu Inches	0.06102
Cu Feet to Cu Metres	0.02831
Cu Metres to Cu Feet	35.3147
Cu Yards to Cu Metres	0.7646
Cu Metres to Cu Yards	1.3079
Cu Inches to Litres	0.1638
Litres to Cu Inches	61.027
Gallons to Litres	4.545
Litres to Gallons	0.22
Grains to Grams	0.0647
Grams to Grains	15.43
Ounces to Grams	28.3495
Grams to Ounces	0.03528
Pounds to Grams	453.592
Grams to Pounds	0.00220
Pounds to Kilograms	0.4536
Kilograms to Pounds	2.2046
Tons to Kilograms	1016.05
Kilograms to Tons	0.0009842

Glossary of Acronyms

AUTOSUB A capability for the routine collection of biological and geophysical data from the deep oceans using unmanned submarines
BBC British Broadcasting Corporation
BIG/ISTIG Biogas Integrated Gasifier/Intercooled Steam Injected Gas Turbine
CFC Chlorofluorocarbon
CFE Conventional Forces in Europe
CIA Central Intelligence Agency
CIG/STIG Coal Integrated Gasifier/Steam Injected Gas Turbine
CLEAR Campaign for Lead-free Air (UK)
CO$_2$ Carbon Dioxide
COBSUB Consumption Obsessed, Supply Based
DDT Dichloro Diphenyl Trichloroethane
DEFENDUS Development Focused, End Use Oriented, Service Directed
DoE Department of Environment (UK)
EEC European Economic Community
FAO UN Food and Agriculture Organization
GCM Global Climate Model
GDP Gross Domestic Product (total value of goods and services produced by a country)
GMO Genetically Manipulated Organisms
IMF International Monetary Fund
IPCC Intergovernmental Panel on Climate Change (a UN established process)
NASA National Aeronautics and Space Administration (USA)
NORAD North American Aerospace Defense Command
OPEC Organisation of Petroleum Exporting Countries
PCB Polychlorinated Biphenyl
PV Photovoltaic cell
RAF Royal Air Force (UK)
RDF Rapid Deployment Force (USA)
SDI Strategic Defense Initiative
UNEP United Nations Environment Programme
UNSNF United Nations Standing Naval Force
USGCRP United States Global Change Research Programme

"When the earth is sick and the animals disappear, the Warriors of the Rainbow will come to protect the wildlife and to heal the earth." In August 1985 the *Rainbow Warrior* was sunk in Auckland Harbour, New Zealand by the French Secret Service.

Introduction

This is a book about security and defence. It is also about unprecedented attacks upon the planet's biosphere, the earth's thin skin which embraces the oceans, land and atmosphere in which all living things have their being. Our subject is environmental security. Security is a friendly, familiar sort of idea. We shall explain that while the old types of threats which we are used to facing, which have enemies behind them, are still present, new kinds of threats – threats without enemies – are crowding in upon us and are likely to be the predominant types of threats in the future.

Indeed, if we don't confront these new threats directly, with meaningful action, and soon, we may find that degradation of the natural environment of which we are a component part and upon which we are wholly dependent, will become a source of conflict in the old fashioned sense, as groups of people fight eco-wars for control of natural resources.

The first part of this book sets out the different ways of looking at "security", sketches out the scope and characteristics of environmental threats and then descends into the abyss, describing the many and powerful forces which block effective protective and remedial action. It tries to imagine the worst of times.

But it doesn't have to be like that. In the second half of the book, we climb out of the pit and describe the many reasons why, given sufficient collective willpower, mankind has a chance to avoid eco-wars and instead, to beat swords into ploughshares in more imaginative and more direct ways than ever before. Furthermore, this is a mission in which each individual can have a role that really counts. Collective willpower is the sum total of all our personal wills, acting in harmony. None of this will happen without the active involvement of each one of us.

Building environmental security calls for an unprecedented act of personal and political mobilization. As a contribution towards that unprecedented act, we have written this book and worked with Lawrence Moore on the television programme of the same name, which is its companion.*

In making this book, we designed the overall structure together, Gwyn Prins then wrote the bulk of the text; Robbie Stamp contributed to several of the technical sections, coordinated and wrote the case studies and prepared other materials from the film.

* *Top Guns and Toxic Whales*, Viewpoint 91, a co-production of Central Television plc, Television for the Environment and the Better World Society. First broadcast on ITV, April 1991. Producer/Director Lawrence Moore, Producer Robbie Stamp, Consultant Gwyn Prins.

Chapter 1: You Can't Shoot an Ozone Hole

The Tomcat and the pram

Dirty weather. Squalls scud across the dark sea. There's a storm rising. The captain has turned the ship into the wind. The big carrier's deck is pitching. But you scarcely notice. Your F-14A Tomcat is fuelled up. You have a full complement of missiles – heat-seekers and radar controlled – on board. You're ready to go. The flight deck crew give you the signal. You gun the throttles to full power and light off the after-burner. The fighter shudders under the thrust. The launch lights go from red to green. The catapult takes hold and you are flung back into your seat under the force of the tremendous acceleration as you are hurled into the sky.

Now you have looked around. You have picked up your wing man and together you streak off towards the horizon. You flick your eye across the displays. Your AIM-7 "Sparrow" air-to-air long range missiles are armed and interactive with the aircraft's fire-control computers. You can engage several targets simultaneously with this system. Suddenly the voice of your controller, seated in an E-2C Hawkeye a hundred miles away, comes through your headphones. "Diamond One, your target is at your nine o'clock, twenty one miles, high and closing."

The briefing on the ship was right. Satellites had given the first warning which had then been confirmed by the airborne early warning radars of the Hawkeye aircraft that was always in the air towards the outer perimeter of the task force's air defence zone. All the different sources of information identifying, tracking and targeting the intruders are operating interactively. Data flows into the missile guidance systems. Computers continuously update your head up display to show you where the enemy is and when he can be engaged. Meanwhile, your navigator is also preparing to defend your aircraft with on-board electronic counter-measures if it is

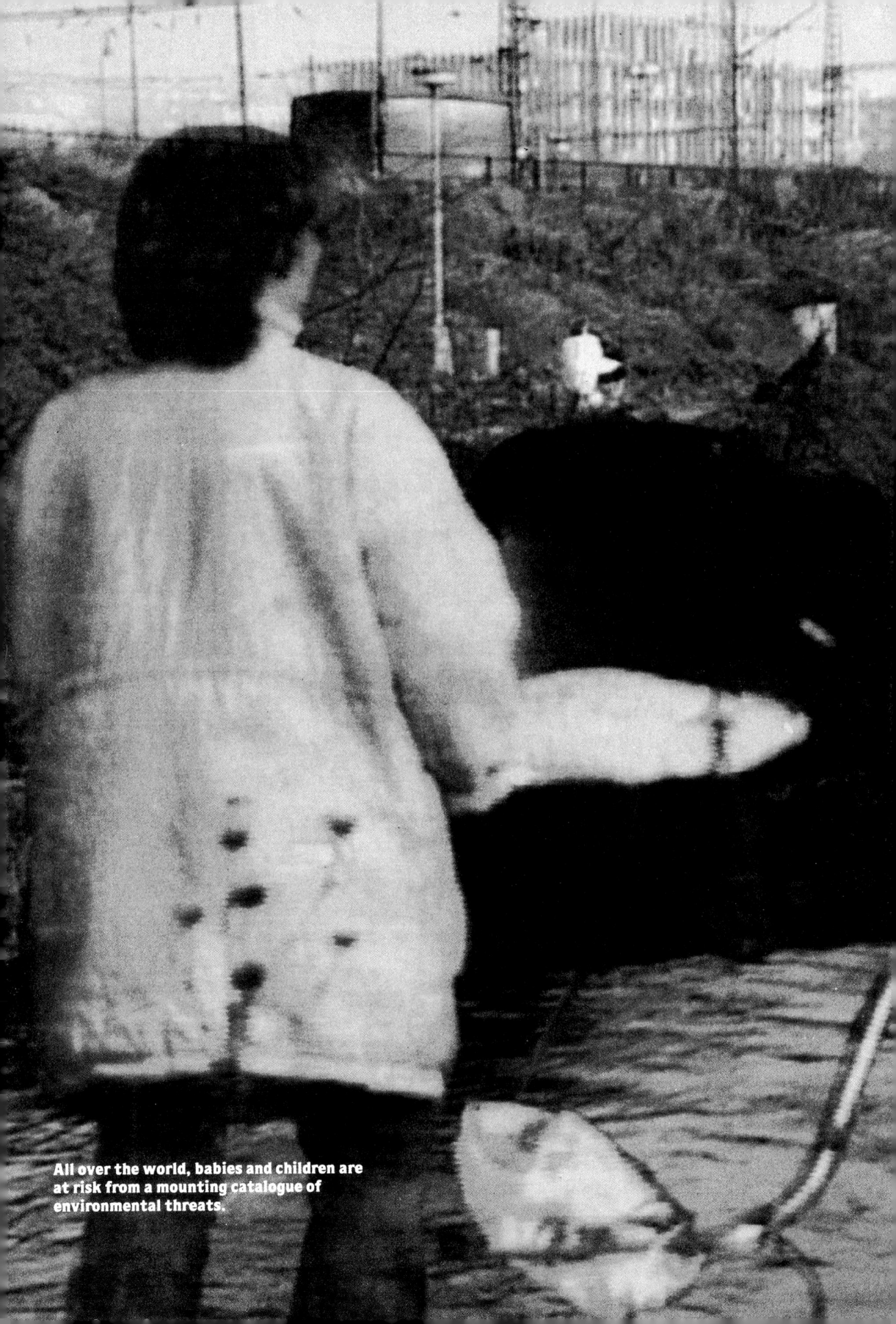

All over the world, babies and children are at risk from a mounting catalogue of environmental threats.

attacked. But much of the success of your mission in attack or defence will rest upon your skill. Because you are Top Gun.

Modern fighter pilots symbolize the pinnacle of the achievement of modern technology applied to the task of identifying a military threat and dealing with it by destroying it. The power and accuracy of modern weapons and their associated sensors is such that often the enemy can be destroyed without ever being seen. This sort of task is something which modern technology can do very well. It is the "sharp end" of the act of defence. Defence against an enemy about whose identity there is no doubt. Your task as a Top Gun fighter pilot is absolutely clear. It is narrowly defined. All your faculties and those of others supporting you in your mission, like all your sensors and weapons, are focused upon a single objective: to destroy the enemy before he destroys you.

If you and all those like you on your side are successful in your missions the sum total of your success is the success of your side against the enemy. That leads to the defeat of the enemy and in his destruction lies your safety. If on the other hand you fail and all those like you fail, then disaster ensues. As the old poem says,

> For want of a nail the shoe was lost
> For want of a shoe the horse was lost
> For want of a horse the rider was lost
> For want of the rider the battle was lost
> – and all for the want of a horse shoe nail.

Now imagine yourself somewhere else. You are feeling very proud but rather tired because the baby in the pram that you are pushing is only four months old. Despite all the exercises and preparations at the antenatal classes, neither you nor your partner had been prepared fully for the momentous experience of the birth when it came.

You've done everything that you possibly can to do the right thing by the baby. Just that afternoon, you carefully read the contents list on the jars of baby food at the supermarket, choosing some for your baby's first taste of solid food after the months of breast-feeding alone. You'd carefully checked and rejected any jars that didn't seem to you to be perfectly pure and uncontaminated with artificial ingredients, even if they were a bit cheaper. And of course your baby will have all her immunizations, all the screenings and tests that the midwife has recommended. She looks so perfect, so peaceful, so completely dependent upon you for food and love and comfort. You wouldn't do anything to harm her.

But what if your pram was seen as a vehicle of defence, upon which the full potential of modern science was lavished, so that it was equipped to the levels of the jet fighter with sensors to detect the threats which lurk around you?

You are walking in a park in Birmingham, England near the famous Spaghetti Junction where three motorways meet. The yellow indicator on your pram handle bar buzzes. The intruder alarm shows up high concentrations of atmospheric lead, an aerosol fall-out from the exhausts of motor vehicles roaring high above you. Studies of blood- and tooth-lead levels in children have shown a generally accepted strong correlation with impaired intelligence. Walking and living here, your little girl is insidiously under attack. But as you walk through the park, if you didn't know that you wouldn't suspect it. It took the Campaign for Lead-Free Air (CLEAR) many years of enduring scorn and calumny from the motor and oil industries, and from government departments, before the break-through came and lead-free petrol became fashionable in Britain, and was given a subsidy to encourage its use.

Next to the lead alarm, another indicator shows up concentrations of carbon monoxide, sulphur, particulates and nitrous oxides in the air. If you were a young Czechoslovak mother walking on a winter's evening, in the heart of Prague, one of the most beautiful cities of Europe, across the Charles Bridge over the Vlatava towards the Mala Strana district, with the magical Hradcany Castle and St Vitus' Cathedral silhouetted by the setting sun, you would not need an air pollution indicator to tell you of the acrid smog, belching from the lignite-burning boilers of the city's houses. The faces on the statues also show signs of erosion by acid rain and bear witness to the threat. If you were walking in Mexico City or Lagos or Beijing or Los Angeles, you might know that you were walking against the advice of the authorities who would have issued a smog or ozone alert. In Los Angeles you would be more likely to drive; and until severe emission controls on automobile exhausts began to be imposed during the last few years, Californian cars were the cause of the grim photochemical smog which could blanket the city under an inversion layer.

Maybe you are lucky enough to live in a city without smog. You are walking in Auckland, New Zealand, for example. It is a bright morning, the air washed by rain the night before. A keen wind is blowing in from the Tasman Sea. You live in a green and sparsely populated, highly developed country, thousands of miles from the smogs and pollution of the northern hemisphere, protected by vast oceans. Surely no problem in this refuge? You push your pram along the waterfront overlooking a spectacular bay. If you had it, modern technology would be ringing another alarm on the handlebar, telling of a threat in many ways more worrying than the smog of Prague or the lead fallout of Birmingham, because much less easily countered. The purple alarm on the pram warns you of excess ultraviolet light, which could easily burn your baby's skin and indeed your own, and if you were exposed to it too much might cause skin cancer or cataracts. Increasing UV in the southern hemisphere is a direct result of the thinning of the ozone layer over the Antarctic as a consequence of the CFC gases from aerosol propellant, refrigerant fluid (Freon) and industrial processes to make fast-food blown-foam cartons.

If you had walked in a London park during the summer of 1990 you might have noticed the sad absence of trees, many of which were blown down in one of the series of "once in a hundred years" storms which have lashed the British Isles over the past few years. In January and February 1990, three "hundred years" storms occurred successively within three weeks. But now it is summer and it is hot! Perhaps, you wondered to yourself, "Is it not really too hot, suspiciously so?" The structural engineer who called last week to look at the cracks which had opened in the walls of your house had said that in the London clay belt, shrinkage had been such that 1990 was proving to be a disastrous year for house subsidence. 1990 was the hottest year in Britain since the sixteenth century and the last four years worldwide have been the hottest years of the century. If you knew that you might have wondered whether you were beginning to see the start of major climatic change.

You are a Canadian and living in Quebec Province. As you push your baby beside the St Lawrence Seaway you might, if you are lucky, see Beluga whales sounding as they swim towards the Atlantic. They at least, you might think, escape the worries of having to live in an apparently normal world where invisible threats assail you and your baby and against which you appear to be defenceless. But you would be wrong. Heavy metals concentrate in the fat layers of mammals, and the whales of the St Lawrence Seaway have such concentrations of pollutants in their blubber that the Quebec authorities have labelled the

Toxic Whales

The Beluga whales in the estuary of the St Lawrence River in north-eastern America are an endangered species. The Belugas are at the top of the aquatic food chain, feeding on eels, herring, capelin and smelt. They live in waters into which drain extensive parts of eastern and central North America. So the health of the Belugas is an index of the health of the St Lawrence system which in turn is an index of the health of the region.

Since 1982, 73 dead Belugas have undergone autopsies. DDT and PCBs were discovered in the blubber at levels which were *among the highest ever recorded in marine animals*. The eels upon which they feed have been found to be especially heavily contaminated with PCBs. Further observations indicate that the DDT, the PCBs and other chemicals such as Mirex are actively transported from the blubber into the internal tissues. So the blubber isn't just an inert reservoir for poisons and the body burden increases toxic stress with age. Baby whales suckling from older females with a high body burden may receive massive transfers of organochlorines and experience toxic shock from their mother's milk.

The St Lawrence Belugas suffer bladder cancer and herpes-like viral dermatitis, discovered in cetaceans for the first time in their case. There is also a significant incidence of gastric ulcers and tumours, extremely rare in studies of other stranded whales.

Regression equation calculations of the total amounts of PCBs, DDT and Mirex presently in the population of 500 whales in the estuary are as follows: 10.09 kg of PCBs, 3.85 kg of DDT and 106 g of Mirex.

Some of these chemicals have not been in use in recent years, but their continuing presence in the bodies of the largest predator in the river estuary demonstrates how pervasive and persistent such poisons are.

It would not be the first time that the fate of whales has been taken as indicative of the larger health of society. The seventeenth-century Dutch were much disturbed by a series of whale beachings on the coast of the Netherlands. They interpreted them allegorically. We must interpret the fate of the St Lawrence Belugas literally.

---

living whales as "toxic waste"; the whales don't know it but their bodies are deemed to be dangerous to human health.

Overhead, the roar of a naval fighter shatters your contemplations as it speeds up river on exercise. But what can a Top Gun do to help a mother worried for the health of her tiny baby? What can a Top Gun do to defend a toxic whale?

Breakdown of the social contract

The state is quite clear about its responsibility to provide Top Gun as part of its wider responsibility to keep its citizens secure, but it seems to be failing in country after country to keep its citizens secure from environmental threat. What is going wrong?

The seventeenth-century political theorist Thomas Hobbes seems to have been a miserable sort of man. Towards his patron the Earl of Devonshire he behaved ingratiatingly, doing his shadier financial business for him in London, being a fixer. To those whom he regarded as his inferiors he showed irritability and suspicion. He was obsessively worried about his health, was a food faddist and a hypochondriac. He lived to be over ninety. But he is mostly remembered for his great book *Leviathan*, published amid much shock in 1651. This analysed the relationship between the individual and the state.

Hobbes's view of society was that people cooperated with each other, observed laws and managed to behave more or less reasonably not out of any innate decency, but principally out of fear. The fear was that if they did not behave, if they stole or murdered, then they would be punished and the punishment would come from Leviathan; so the fear was a good thing. "During the time men live without a common power to keep them in awe, they are in that condition which is called war," he wrote, "and such a war is of every man against every man.... The nature of war," he explained, "consisteth not in actual fighting but in the known disposition thereto during all the time there is no assurance to the contrary." In other words, Hobbes said that the menace hanging over you before the sword was drawn was just as much a state of war.

Leviathan would express the voice of the community and would impose punishment so severe that the individual was deterred from antisocial actions and a semblance of order in society thus maintained. Without this discipline, "No arts; no letters, no society; and what is worst of all, continual fear and danger of violent death." Men's lives would be, Hobbes famously observed, "solitary, poor, nasty, brutish and short".

The eighteenth-century French philosopher Jean Jacques Rousseau had a much less bleak view of human nature than Hobbes. He lived at the time of the Enlightenment, in which Europeans experienced the great optimism which came from the recognition that science and technology were placing in their hands the power of knowledge and the power of invention to impose Man's will upon Nature and to drive back the frontiers of the Wild. It was also a time when reason and science were seen to be allied against privilege and superstition. *Liberté, égalité, fraternité* was the great slogan of the French Revolution – which nevertheless produced the Terror and Napoleon. In his book *The Social Contract* Rousseau argued that order and discipline in society, acting to the benefit of the individual although constraining the individual's absolute freedom of action, could be achieved by the operation of man's higher and more generous faculties. It was logical and sensible for people to agree together not to do certain things (such as to murder each other or not to pay their taxes) since they could freely accept that making such a contract with people collectively (the state) had advantages for all: it gave a guarantee of predictable behaviour and allowed a definition of "normality".

So Hobbes in his baleful manner, and Rousseau, more optimistically, described the same thing: the individual would give up some of his or her putative absolute freedom for the gain of the greater freedom that came from agreeing a social contract. But why make the contract with Leviathan, with the state? For an obvious reason: because the state was all-powerful. It was possible either to bribe the state (Hobbes's view) or to persuade the state (Rousseau's view) to give you, the individual, this contract. And hypothetically, if the individual was prepared to give up enough, the state was in a position to deliver absolute satisfaction; because all power lay in the hands of Leviathan.

Under this contract, if the individual broke his or her part, then the retribution of the law, or of social ostracism, or of bankruptcy might follow. If, on the other hand, the state broke its part of the contract by abusing the human rights of the individual, then the individual had a legitimate right of rebellion to overthrow the unjust state. Europe has recently seen an enormous and encouraging example of individuals joining together to overthrow illegitimate states in the 1989 revolutions which swept across the eastern part of the continent, breaking down the Berlin Wall, shattering the Iron Curtain and reuniting the two halves of the continent split apart at the end of the Second World War.

This basic view of the contract between the individual and the state gave rise to the first level of definition of the state's responsibility to the individual. This is often called the responsibility of the "night watchman state". The night watchman state has the responsibility of defending the individual physically from aggression. In order to do this the state may legit--

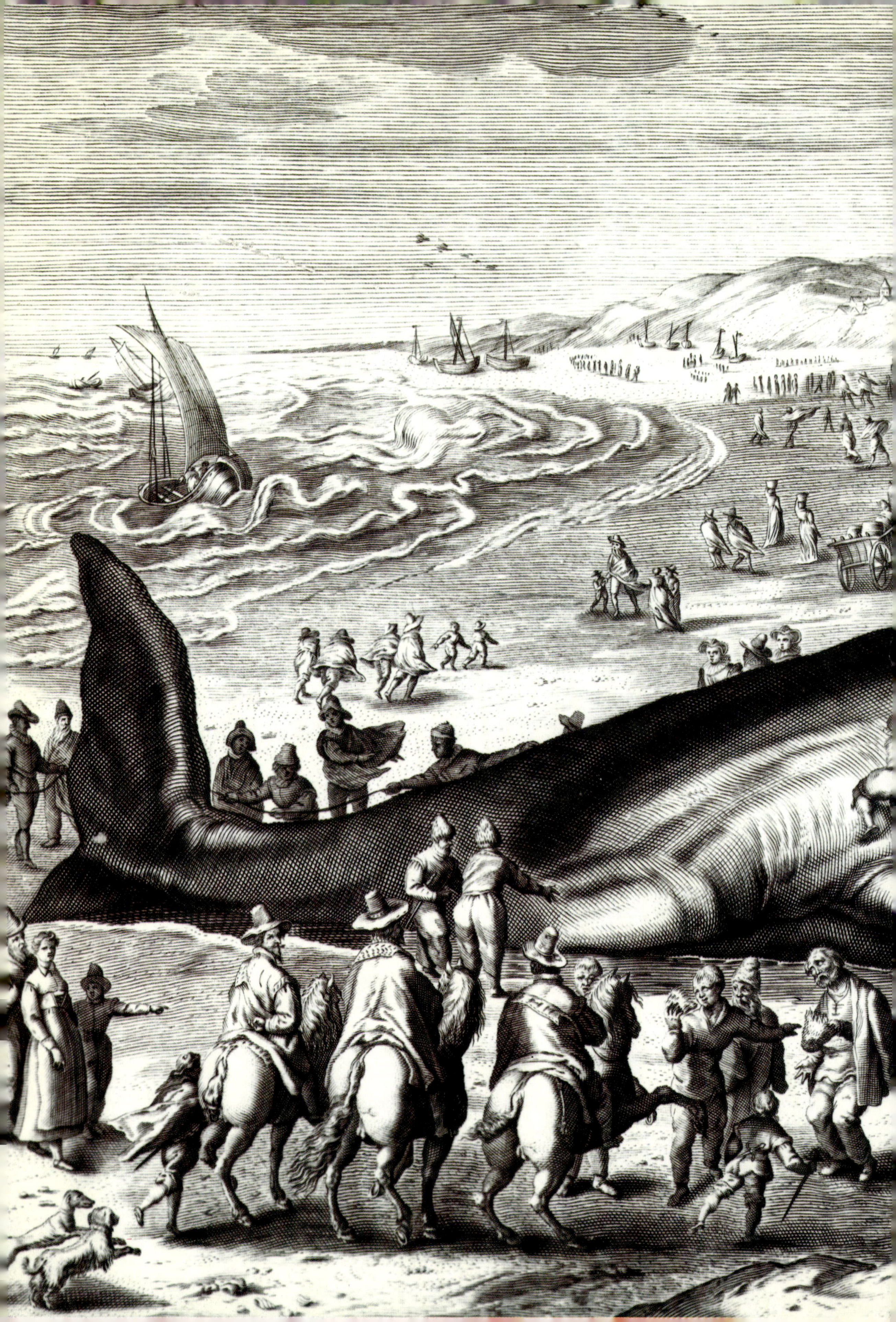

In the sixteenth century, ill health was interpreted allegorically; we must interpret the fate of these whales literally.

imately raise taxes to pay for armed forces to defend the individuals within the state from the enemies outside its borders, and to recruit policemen in order to maintain law and order and thus to defend the individual from enemies within. This rather spartan view of what the state was obliged to do under the contract remained its dominant interpretation until the middle of the twentieth century.

Then the view took hold that the minimum responsibilities of the state towards the individual were larger than this. The individual had the right to expect that the state would guarantee his or her access to basic facilities such as health, education, perhaps also employment and if not employment, unemployment benefit. For many people in the northern industrial countries, the willingness to embrace the concept of the welfare state in the 1940s was in large part a reaction to the misery that individuals had experienced during the depressions of the 1930s when state support for people in times of economic and social distress had been much less. These wider objectives were expressed formally in many places including in the wording of the Charter of the United Nations in June 1945.

Since then, for a generation, domestic politics in many parts of the world have revolved around the question of the distribution of wealth. How much should people be compelled by taxation to give up of their earnings so that this surplus may be redistributed in various forms to others who did not earn as much? To what extent should people be allowed to pass on without interference their lifetimes' savings to their children? But what was never in doubt in the debate between socialism and a more austere political philosophy was the ability of the state to deliver these services and goods should the will exist to make it do so.

This in turn was grounded in a fundamental assumption of modern society since the eighteenth century Enlightenment: that there is an essentially limitless world from which mankind can extract raw materials. These, by ingenuity, we can manufacture into the whole range of things necessary to make modern civilized life and when those things wear out we can safely throw them again back into the vastness of Nature and forget about them. At this stage, the biosphere was not in question. It was just part of the background against which the human drama was being played out. The solution to pollution was dilution. An easy and reassuring refrain.

> "We, the peoples of the United Nations, determined to save succeeding generations from the scourge of war, which twice in our lifetime has brought untold suffering to mankind and . . . to promote social progress and better standards of living in larger freedom . . ." (Preamble to the UN Charter)
>
> "With a view to the creation of conditions of stability and well-being which are necessary for peaceful and friendly relations among nations . . . the United Nations shall promote: (a) higher standards of living, full employment, and conditions of economic and social progress and development (b) solutions of international economic, social, health and related problems; and international cultural and educational cooperation" (Article 55 of the UN Charter)

If the biosphere was thought about at all, it was as a tolerant dustbin; the possible circular routes through food chains, and through the cycles of the elements through the biosphere, were discounted in the operating decisions of day-to-day life. It is a basic difference. The reality of the natural world is that its processes are circular and all things are connected to all other things; the power of modern thought has been its ability to unpick complicated things and to look at each component part in isolation, to view processes as linear with a beginning, a middle and an end. It is a deadly illusion.

For the worried mother walking in the park, pushing her pram, and for the rest of us, the illusion can be no longer entertained. Under the old way of thinking, "defence" was more or less synonymous with "security", and security was keeping out or keeping under control those elements which might disrupt normal peaceful predictable life. In the face of environmental threats of the severity which face us all today, security must be differently defined. In what way?

Redefining Security

Under the old regime, security was about keeping people who were not like us, or whom we did not like, out. It was about living with and having a sense of community with people like us; it was also about minimizing the unexpected: maximizing predictability. This was not only the predictability that we expect in our social contract with others and with the state. Predictability is also something which the rich have come to expect in their continuing access to a very wide spectrum of what they perceive to be the "necessities" of everyday life. That list now extends far beyond the basic requirements of food and shelter and clothing and warmth. It includes "rights" to transport, to information, to a vast range of consumer goods, to constant electric power. The Founding Fathers of the American Republic in the Declaration of Independence had a list of three things: life, liberty and the pursuit of happiness. They came in that order for a rather important and obvious reason. Without life, you can do nothing else; without liberty, you cannot have a fulfilled life; and the pursuit of happiness (interpreted in the material sense) is the cream on the top. Many modern people in industrial societies, however, have confused the priorities and have come to see their rights to the pursuit of happiness to be more important than other peoples' rights to life or to liberty.

So this is a definition of security which is *exclusive*. We don't see our security as being bound up with other, very different peoples' security. If we think about it at all, we accept that our security could co-exist with – maybe even be a product of – their insecurity: "If the poor starve, it's their own fault for being so lazy/stupid etc., etc. In this life it's survival of the fittest. The race goes to the strong." That exclusivity relates to a very important feature of the identity of modern peoples and to the nature of the contract which we have made in modern societies between ourselves and with the state.

Our identity in the "nation states" in which we live is tightly bound up with a whole bundle of things. "National Security" has come to be seen as really the same thing as the security of the individual living within that particular state. So the individual's security becomes part of maintaining the territorial integrity of the state or indeed – as was the case during the East-West Cold War – the maintenance of the ideological supremacy of a set of ideas ("communism", "capitalism", or whatever) which that state is said to represent.

As the appetite for raw materials and also for markets in these very wealthy states grew during the twentieth century, so did these same states' conception of what their minimum security was. It came to involve their right not only to territorial integrity but also to access to raw materials that came from other parts of the world at fixed and stable prices. Indeed, the great American economist John Kenneth Galbraith has argued that once an individual or a company or a country has reached a certain point of wealth their economic rationale comes to be less and less driven by the pursuit of profit, more and more by the pursuit of predictability and control of

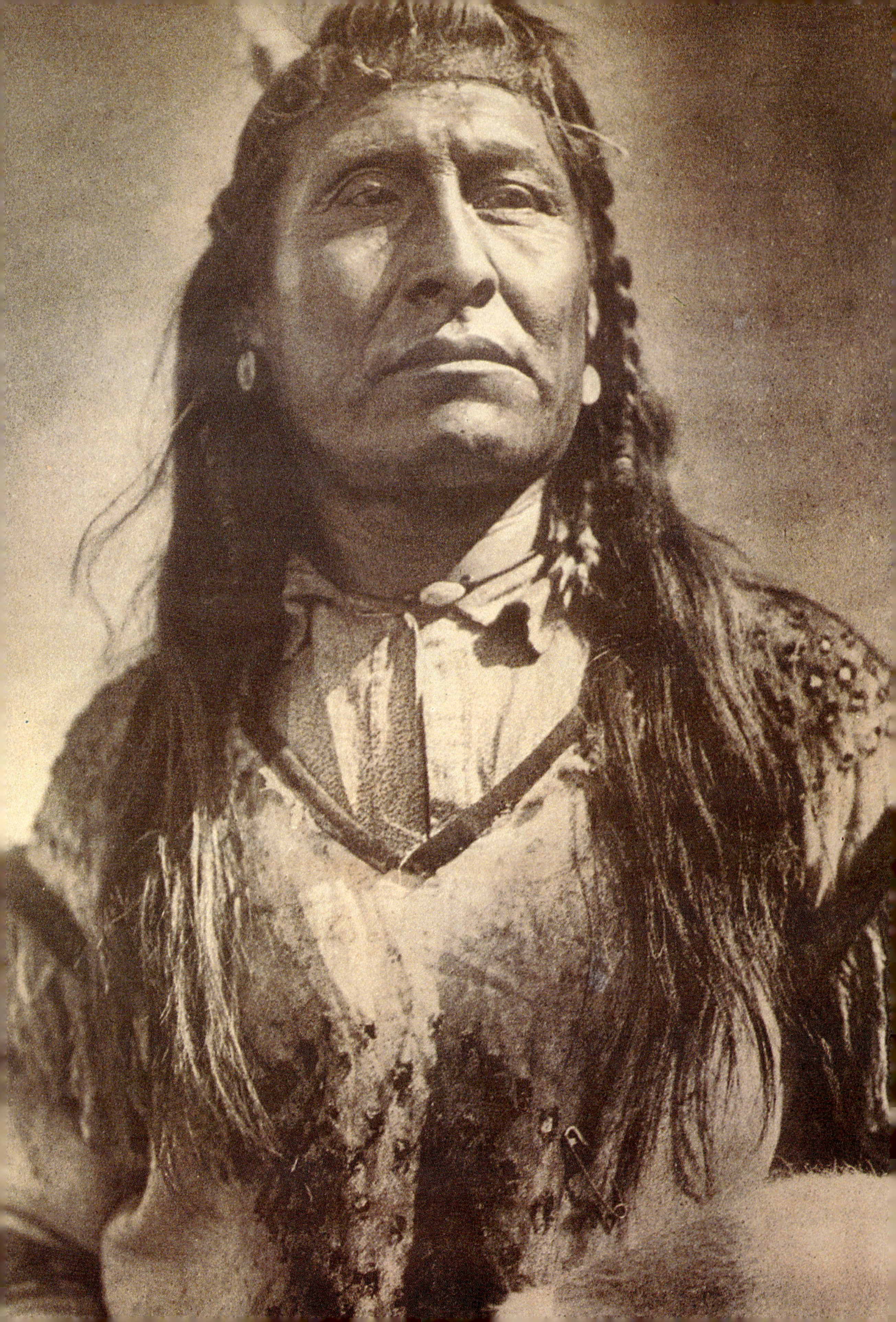

Edward Curtis took these photographs at the turn of the century.

their environment. So companies come to be more interested in controlling each part of the process in which they are involved – from raw material to manufacture to market research to marketing – than in the simple profitability of their particular product.

These enormous transnational companies dominate important areas of the world's economy: oil, basic foodstuffs, mass-produced goods, electronics. Their growth was one of the first signals that the power of the nation-state, previously presumed to be absolute, was beginning to be seriously constrained. Workers found this out soon enough and bitterly enough. If they tried to strike for better wages from the Ford Motor Company then the Ford Motor Company could simply transfer production to another of its plants in another country with a different labour force and in this way the power of the worker as against these enormous conglomerate institutions was mightily reduced.

This coincided with the dramatic demonstration in the years after the Second World War of the *decreasing* significance of military power as the primary expression of the virility of a state and the *increasing* significance of economic power as a primary expression of influence in the international community. How the power of multinational enterprises is to be made responsive to public will is still an unsolved problem. In a crude sense, the market can dictate; but, as Galbraith has explained, such companies take special care to insulate themselves from market forces. They act like nation-states but are not amenable to the mechanisms of the United Nations, for example, which are framed upon the presumption that the state is the focus of the central powers which must be made internationally accountable. Article 59 of the UN Charter contains the only reference to a possible mechanism for control, but its tentativeness is patently incommensurate

> **The Organization shall, where appropriate, initiate negotiations among the states concerned for the creation of any new, specialized agencies required for the accomplishment of the purposes set forth in Article 55 [see p.6 above]. (Article 59 of the UN Charter)**

to the task.

In this new world of an internationalized economy which grew during the "long boom" and which lasted from the end of the Second World War to the oil-price shocks of the early 1970s, a period of cheap primary energy for the industrial world, people still did not recognize a new and active player who demanded to be considered in the maintenance of predictability and control in the domestic environment. This stranger was the biosphere.

In the 1960s the first astronauts came back to earth, bringing with them photographs of an arresting beauty. They showed the blue planet Earth set against the pitch black of outer space. For the first time, millions of people could experience not only that sense of wonderment which the astronauts themselves had felt but also the shocking realization that the earth, which we had to all intents and purposes in daily living regarded as an inexhaustible resource, was not only finite but small and fragile, set against the backdrop of limitless space. Furthermore, in comparison with the desolate barrenness of all other heavenly bodies, the earth was visibly pulsating with life. Indeed, from a distance it looked as if it was a single living organism.

That the earth is a living creature is scarcely a novel idea. The Ancient Greeks deified the earth, revering her as the earth goddess, Gaia. Gaia was responsible for the wellbeing of her domain. She would reward those who cared for her and punish through the agency of natural catastrophes – through drought, famine, earthquake, storm and plague – those who destroyed her forests, or eroded her hills or killed her creatures. Many peoples expressed such a view in animist religions, religions which worshipped spirits in trees, in places, in stones and hills. It was a view of humanity's relationship to the earth which came in the end into direct conflict with a different attitude. In contrast to that sense of the interconnectedness of all things, the later attitude assumed a great gulf between human beings and the rest of the natural world; saw them as different things. The natural world was seen now not as the abode of spirits but as a resource to be harvested.

In this confrontation, the old view almost invariably lost out. Faced with the necessity and inevitability of surrendering his lands to the white man, Chief Seattle of the Pacific Northwest of the United States made a famous analysis of what it was that distinguished the red men and the white men.

Both Chief Seattle's testimony and Edward Curtis's photographs of the last of Indian life lay buried and forgotten for many years. But they were both rediscovered in recent years as an increasing awareness took hold that the existing definition of security was insufficient. Its narrow sense, the "defence" definition, bounded by ideological conflict in the Cold War and the narrow and selfish interests of nation states could not cope; nor could the wider but still traditional sense of economic well-being of the state or of welfare for the individual within the state.

Historically the first signal of the new problems emerging on to the scene is associated with the publication in 1962 by Rachel Carson of *Silent Spring*, her famous study of the reckless use of pesticides and herbicides. *Silent Spring* told the dreadful story of the destruction of wildlife and of wild lands in the United States, and of the effects upon human health of the indiscriminate and ignorant use of agricultural chemicals, notably DDT to kill insects and 2,4-D to kill broadleafed plants. The voice of Chief Seattle could be heard through hers.

> "How can you buy or sell the sky? We do not own the freshness of the air or the sparkle of the water. How then can you buy them from us? Every part of the Earth is sacred to my people, holy in their memory and experience. We know that the white man does not understand our ways. He is a stranger who comes in the night and takes from the land whatever he needs. The Earth is not his friend but his enemy, and when he has conquered it he moves on. He kidnaps the Earth from his children. His appetite will devour the Earth and leave behind a desert. If all the beasts were gone, we would die from a great loneliness of the spirit, for whatever happens to the beasts, happens also to us. All things are connected. Whatever befalls the Earth, befalls the children of the earth."
> (Chief Seattle)

People began to worry about pollution. People began to worry that the world was sick. Many found solace in recovering that sense of mystic wholeness which had marked the animist religions.

But science itself began to see things in a different way and to develop philosophically beyond the Neanderthal stage which Rachel Carson castigated. Developing slowly over thirty years and bursting upon the political scene within the last decade, a quiet revolution in the application of the scientific method to the study of the natural world has taken place.

The romantic poet William Wordsworth wrote against the scientific method at the beginning of its rise to power.

Our meddling intellect
Misshapes the beauteous forms of things:
we murder to dissect.

Enough of science and of art:
Close up these barren leaves.
Come forth, and bring with you a heart
That watches and receives

> As man proceeds towards his announced goal of the conquest of Nature, he has written a depressing record of destruction, directed not only against the earth he inhabits but against the life that shares it with him. The chemical weed-killers are a bright new toy. They work in a spectacular way; they give a giddy sense of power over nature to those who wield them, and as for the long-range and less obvious effects – these are easily brushed aside as the baseless imaginings of pessimists. As crude a weapon as the cave man's club, the chemical barrage has been hurled against the fabric of life . . . this birth-to-death contact with dangerous chemicals may in the end prove disastrous. . . . As matters now stand, we are in little better position than the guests of the Borgias. . . . The control of Nature is a phrase conceived in arrogance, born of the Neanderthal age of biology and philosophy, when it was supposed that nature exists for the convenience of man. . . . It is our alarming misfortune that so primitive a science has armed itself with the most modern and terrible weapons."
> (Rachel Carson, *Silent Spring*)

But it is *only* by dissection, by working out the basic principles of what makes things work, that we advance in systematic understanding. The tragedy of a scientific approach divorced from a wider ethic was that dissecting did indeed murder; yet Wordsworth's stark polarity is not inevitable. That is what is at issue, for example, in the difficult argument about medically necessary and commercially motivated vivisection between scientists, cosmetic companies and animal rights campaigners. It has been in the science of ecology that the dissecting necessities of the scientific approach have been transcended and a new level of understanding, which can recognize and take account of the circularity of processes and the interconnectedness of the fabric of the biosphere, begins to be achieved.

One of the many who looked with wonderment at the astronauts' photographs of the earth from outer space was the scientist James Lovelock. To him, the whole phenomenon seemed strange. Why should this blue and living earth exist, a chemical-powered anomaly in a generally nuclear-powered universe? How had life originated on this planet; how had the planet managed to remain comfortable for life in the face of so very many fluctuations in its cosmic history? Lovelock had had experience working on atmospheric chemistry and so for him the most astonishing thing about the earth's atmosphere was that it seemed to have remained remarkably stable for long periods in the history of the planet, periods in which the output of energy from the sun may have varied by as much as 30 per cent.

After the greatest "pollution crisis" in the planet's history, when oxygen increased in the atmosphere to levels which made the surface uninhabitable by the anaerobic life forms which had previously dominated it, the earth's atmosphere had maintained its composition: mostly nitrogen, about 21 per cent oxygen and less than one per cent carbon dioxide. This, Lovelock reasoned, was not an atmosphere in steady-state equilibrium. Equilibrium would be like the composition of the atmosphere of Mars: 96.5 per cent carbon dioxide, less than 1.8 per cent nitrogen and less than 0.1 per cent oxygen. It would be an atmosphere without potential, for all, or virtually all, chemical reactions within would have gone to completion: Warm, damp and lifeless. In contrast, the Earth's atmosphere is full of chemical power: oxygen at a level only 4 per cent below that which would make the whole atmosphere inflammable means that almost anywhere, fire can be lit and chemical-free energy harnessed.

No, the earth's atmosphere was in a stability which, Lovelock believed, was not being maintained by accident. He concluded that the Earth's atmosphere was not just a geophysical or even a biological *product*, it was in fact a biological *construction*: kept comfortable for life by

A planet fit for life

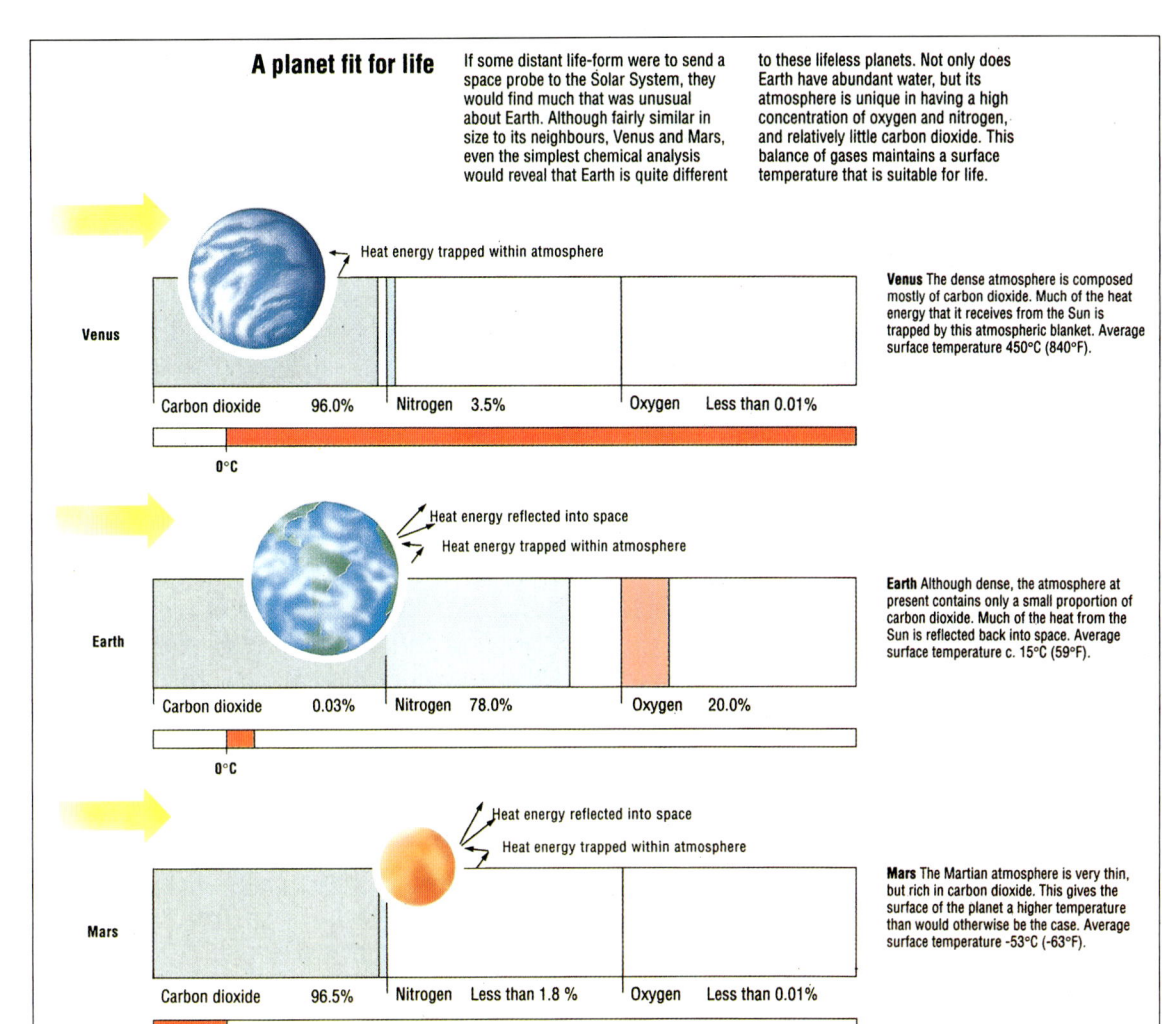

If some distant life-form were to send a space probe to the Solar System, they would find much that was unusual about Earth. Although fairly similar in size to its neighbours, Venus and Mars, even the simplest chemical analysis would reveal that Earth is quite different to these lifeless planets. Not only does Earth have abundant water, but its atmosphere is unique in having a high concentration of oxygen and nitrogen, and relatively little carbon dioxide. This balance of gases maintains a surface temperature that is suitable for life.

Venus

Heat energy trapped within atmosphere

Carbon dioxide 96.0%   Nitrogen 3.5%   Oxygen Less than 0.01%

0°C

Venus The dense atmosphere is composed mostly of carbon dioxide. Much of the heat energy that it receives from the Sun is trapped by this atmospheric blanket. Average surface temperature 450°C (840°F).

Earth

Heat energy reflected into space
Heat energy trapped within atmosphere

Carbon dioxide 0.03%   Nitrogen 78.0%   Oxygen 20.0%

0°C

Earth Although dense, the atmosphere at present contains only a small proportion of carbon dioxide. Much of the heat from the Sun is reflected back into space. Average surface temperature c. 15°C (59°F).

Mars

Heat energy reflected into space
Heat energy trapped within atmosphere

Carbon dioxide 96.5%   Nitrogen Less than 1.8%   Oxygen Less than 0.01%

0°C

Mars The Martian atmosphere is very thin, but rich in carbon dioxide. This gives the surface of the planet a higher temperature than would otherwise be the case. Average surface temperature -53°C (-63°F).

control mechanisms driven by life forms dependent upon the maintenance of this dynamic stability and latent chemical potential for their continuance. The fundamental virtuous circle.

> Having worked on the Martian atmosphere, looking for signs of life, I later switched back to earth and concentrated on the nature of our own atmosphere. This work resulted in the Gaia hypothesis, which suggests that the entire range of living matter on earth, from whales to viruses, can be regarded as a single entity, capable of maintaining its environment to suit its needs. . . .
>
> We defined Gaia as a complex entity involving the earth's biosphere, atmosphere, oceans and soil, the totality constituting a "feedback" or "cybernetic system" which seeks an optimal physical and chemical environment for life on this planet. Gaia remains a hypothesis, but much evidence suggests that many elements of this system act as the hypothesis predicts. (James Lovelock in *The Gaia Atlas of Planet Management*)

In Lovelock's hypothesis there emerged, for the first time since Darwin's *Origin of Species*, a scientific idea which has as its consequence a fundamental recasting of our image of the world in which we live. Suddenly we see a possible coming together of the mystical and the scientific strands of human understanding. For if it is indeed the case that the relationship between the biosphere and the atmosphere is most usefully thought of through the analogy of a living organism, then the hitherto unbridgeable gap – between the meddling intellect which murdered to dissect and a heart that watches and receives – can finally be overcome. An important part of the present challenge to define global security is at this philosophical level and aims to reconcile the two.

The threats are increasing, for mankind has not been inactive. We have increased the volume through-put of the carbon cycle by 20 per cent, the nitrogen cycle by 50 per cent and the sulphur cycle by 100 per cent. We have poured toxins into the air, the water and food chains. We have reduced the green cover of the planet and as the number of people grows, so do the accompanying stresses upon the environment. Just as the concept of security had to acquire economic and social dimensions a generation ago, in the late 1980s and 1990s it must now acquire an ecological and environmental dimension.

We have argued that security is about things that link together the individual and the group. Once, security was a concept defined at the national and international scale above all. Now we add both the smaller scale – personal security – and the larger scale – environmental security. The fact of the close coupling of these two, the very small and the very large, is one of the distinguishing characteristics of global security.

A question comes from the mother pushing the high-tech pram and an observation comes from the Top Gun pilot. "As you tried to imagine yourself into my position," the mother asks, "visualizing all the many threats that did not have identifiable enemies identifiable behind them but which bore in upon my baby, would *you*, knowing what the warning indicators told you, put up with it?"

Top Gun observes that those threats which do not arise from identifiable enemies cannot be dealt with by his guns and missiles. Introduction of the environmental dimension into the concept of security could not come at a more timely moment; for the inherited concept of security from the last generation has reached a moment of crisis.

Leviathan meets Humpty Dumpty

The Second World War ended with the dropping of the first atomic bombs upon the Japanese cities of Hiroshima and Nagasaki and ushered in the nuclear age.

The atomic bombs were followed in quick measure by the infinitely more powerful hydrogen bombs. How on earth were strategic thought and international relations to accommodate this monstrous increase in the destructiveness of war?

At the same time that the atomic bomb arrived on the international scene, so did the Cold War. The former allies who together had combined to defeat Hitler now found themselves staring at each other across a divided Europe. The West found a way out. It was recorded in 1950 in a famous document written by Paul Nitze. The charter document of the nuclear age has an unprepossessing title: *NSC-68*. What *NSC-68* said was that since the Soviet Union was a state unlike normal states one could not do normal business with it. It could not be negotiated with, it could not be trusted because it was a supremely ideologically based state and its ideology was irreducibly expansionist. The only thing that could be done with countries like this, argued *NSC-68*, was to encircle them with a ring of steel, to "contain" them. (An interesting footnote in the history of the Cold War is that when Stalin was given a report on *NSC-68* the word "contain" had been rendered as "strangle".)

How then was this pariah state to be contained and deterred? The answer was equally obvious. At just the right moment, the immense power of the atomic bomb had arrived and so at a stroke the problem of the Russians could be solved and the problem of the bomb could be solved by linking the two together. The bomb could be used to deter the Russians from the expansionism which they were presumed to wish to undertake.

The result of the adoption of this analysis as the underlying intellectual framework of the next thirty years was to militarize all aspects of the analysis of international affairs. Whatever it was that made the Russians and Americans differ-

> **The Soviet Union, unlike previous aspirants to hegemony, is animated by a new fanatical faith, antithetical to our own, and seeks to impose its absolute authority over the rest of the world. Being a totalitarian dictatorship, the Kremlin's objectives in these policies is *(sic)* the total subjective submission of the peoples now under its control. The concentration camp is the prototype of the society which these policies are designed to achieve, a society in which the personality of the individual is so broken and perverted that he participates affirmatively in his own degradation. The Kremlin is inescapably militant. It is inescapably militant because it possesses and is possessed by a worldwide revolutionary movement, because it is the inheritor of Russian imperialism, and because it is a totalitarian dictatorship. . . . It is quite clear from Soviet theory and practice that the Kremlin seeks to bring the free world under its dominion by the methods of the Cold War. (Paul Nitze, *NSC-68*)**

ent from each other could be now ultimately reduced to an abstruse calculation of the mathematics of Armageddon. In 1963, shortly before his assassination, President Kennedy made a speech at the American University in Washington in which he invited Americans to try just for one moment to consider what the world might look like through Russian eyes. This was regarded with shock by many in the American strategic community.

There lay a lesson in the shock, because it showed to what degree the use of such powerful images as Nitze had conjured up in *NSC-68* had served to distance, dehumanize and indeed demonize the potential enemy. Many who have studied the psychology of the nuclear age have argued that such demonization and distancing is indeed a necessary prerequisite for people who wish to retain their sanity while at the same time planning the nuclear incineration of hundreds of thousands of people.

The extreme militarization of thinking and the demonization of one's opponents that were the mark of Cold War thinking have also other qualities. These are more hopeful for our tasks of the moment, because they explain how quickly attitudes can change. In common with other forms of fundamentalist or extreme views, the security concept of the Cold War world, which hogged all our attention and all the resources at the time that it held sway, had a hidden weak point. It had about it a brittleness which meant that as soon as any one of its fundamental struts could be shown to have been dislodged the entire structure could collapse.

In fact, public opinion research reveals that in 1989 and 1990 that is precisely what happened. An entire complex of views about enemies and Russians which had been held with considerable stability since 1946 by the western public could be seen in the opinion polls to be suddenly, dramatically disintegrating. The pattern of this collapse looks remarkably similar to that which was documented by Mass Observation – the pioneer public opinion research organization – at the end of the Second World War. On that occasion the popular image of Germans as people who were indescribably evil, shifty, untrustworthy and hideous was transferred, lock, stock and barrel, in the minds of the western public, from one set of enemies, the Germans, to another set of enemies, our former friends, the stout hearted Soviets, subjects of genial Uncle Joe.

On 2 August 1990 President Saddam Hussein of Iraq invaded the neighbouring country of Kuwait. The response was the most tremendous mobilization of military power outside Europe to have occurred since the end of the Second World War. The danger is that at just the time that the old enemy image of the Russians was cracking apart Muslims, particularly fundamentalist Muslims, as a result of these actions in the Middle East may become the unfortunate new recipients of that bundle of enemy images that shifted first from the Germans to the Russians and seeks now to shift from the Russians to somebody else. Public opinion in the West has, for the last year and a half already, been indicating with increasing strength that the enemies which it fears most are terrorists who blow aircraft out of the sky, drug runners, "fundamentalists" and religious fanatics, in short anybody who seems likely to pose a threat to the "normality" of life in the rich northern hemisphere. Of course, it is perfectly reasonable to abhor the actions of many of these groups. The trick is not to fall into the trap of demonizing the other party: to do this is to condemn oneself never to find a permanent solution to the problem which the other party represents for you.

Needless to say, seen from the perspective of the poor South – from Bangladesh, Bolivia or Botswana – the range of threats to security is importantly different. In many parts of the poor world, the source of instability and insecurity arises from the fact that the social contract between the individual and the state is not now, and in many places never has been, secure. The poor world is riven by civil war and it is a terrible tendency of such internecine conflicts that they often come on top of or are amplified by other stresses that individuals experience as a result of living in areas where the climate or the soil gives them a hard living. In Liberia, Ethiopia, the Sudan, the Philippines, Central America, famine is a weapon in civil war. War, famine and pestilence – the horsemen of the apocalypse – tend to ride together.

So the challenge is bigger still. It is not just to create a new and more realistic concept of security, one which recognizes that the realm of the domestic now embraces the entire planet because we have to take care of the biosphere just as if it were our own back garden. It comes exactly at the moment of the cracking apart of the brittle eggshell of the security con-

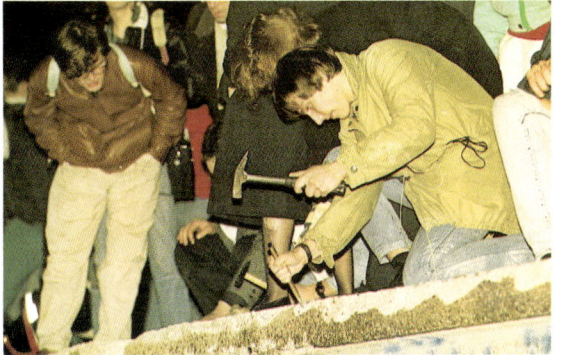

Top: Churchill, Roosevelt and Stalin at Yalta.
Above: The Berlin Wall finally came down on 9th November 1989: people took hammers to the central symbol of the Cold War.

cepts of the Cold War age. But it comes also at a time when there is great danger that those old habits of mind may easily and damagingly redirect themselves from an East–West to a North–South axis. Nothing could be more dangerous to the interests of North–South relations in the old-fashioned and conventional sense, or of global security in the emerging sense, than to allow such a reorientation to happen. Humpty Dumpty has fallen off the wall and it is very important to ensure that all the king's horses and all the king's men can put Humpty back together again.

The shocking realization that comes with the arrival of global security at the centre of the world's political agenda is that the day of Leviathan's all-powerful reign is clearly over. For no single state can guarantee personal security of the sort that the young mother so urgently desires for her child and for her family in the face of the invisible threats which come from every direction and from none, on the wind, in the water, in the food, in the soil. It is in the damage to the environment that these global threats are first being seen and feared.

Ruth Leger Sivard, whose *World Military and Social Expenditures* year by year compares the way that the world uses its resources, described the consequence of the nuclear security policies of the last generation in terms equally applicable to the environmental security problems of the next:

> Every hamlet has been brought within the orbit of conflict, every inhabitant made a potential victim of random annihilation. Militarization presumably designed to insulate and protect the nation state has in fact united the world's population in a precarious mutual vulnerability.

Who then can offer security in an age when a single state cannot? In the short term, paradoxically, the answer has to be – the nation state. But the nation state acting now in a community, the different members overriding their divisive and destructive differences and finding ways to use the supra-national organizations of states, preeminently the United Nations, to harness their collective efforts to the new collective cause. Nor is it just that the nation state is still powerful in the material world. It is also powerful in the minds of its citizens and, as of yet, there is no organizing agency that is sufficiently widely endorsed to substitute for it.

We face now the task of identifying a new enemy far more elusive than one that can be rendered in the form of particular countries or particular peoples. For the source of these threats lies in enemies that include ourselves.

Does it also follow that the *processes of threat analysis* developed in the previous age are also of no help to us? No. Field Marshal Montgomery of Alamein used to teach the principles of warfare to young officers. "The first principle," he said, "is to identify your enemy accurately. The second principle is, having identified him, maintain your aim!"

So, following Monty's advice, let's go into the Situation Room and identify our enemies. The Environmental Security Situation Room has a code name, like all such places. It is called CASSANDRA. In the next chapter we will put the enemies up on the screens in CASSANDRA. Then, in Chapter 3, we shall argue that there are, in fact, useful things that we can take with us into the battle for environmental security from the means of analysing threats made familiar by military men engaged in defence.

An ancient North American legend predicted that, "When the earth is sick and the animals disappear, the Warriors of the Rainbow will come to protect the wild life and to heal the earth."

The full spectrum of the challenge of

environmental security means recognizing and accepting that the ways of thinking and the powers of the nation state will not be able to offer an analysis or a long-term action plan to defend the mother pushing her pram. The Top Gun is powerless to protect her. There are no targets for him to shoot at. But the nation-state remains powerful in the present, and in the minds of people, and this is a battle from which no one is excused. Thus the challenge we face is to use all our means to find those parts of the inheritance from the old security definitions which can be carried with us and employed usefully in the redefinition of security which is now being undertaken. It is a deep paradox.

If we fail to redefine security successfully, we may find that we live in the worst of times, and will have to call in the future upon the traditional military response, resolving nothing, in a succession of battles over access to water and to food as environmental degradation bites, like the more familiar battles over access to and control over raw material, familiar from the past and present. To avoid this grim prospect, we must mobilize and redirect conventional contemporary security concepts to new tasks.

Chapter 2: Visiting CASSANDRA

The War Room

It's very quiet in this large, darkened room. Only the whisper of the air conditioning and a low hum from the machines. The place looks like NASA Mission Control in Houston, Texas, where we watched NASA talking to the astronauts. It is the kind of room in which Peter Sellers portrayed the deadpan president, the neurotic RAF exchange officer and the demonic scientist all at once in Stanley Kubrick's famous film *Dr Strangelove* or, from another film, *Wargames*, the room in which a maverick computer threatens to plunge the planet into World War Three.

There are dozens of computer terminals in rows on desks which form a series of concentric semicircles. The room is built in the round, so that the operators can all have a clear view of the screens. Many screens run round the walls. Although there is quite a lot of light from the screens and from the maps ranged along the walls showing areas of special strategic significance, the impression is of a cathedral with figures seated in their pews or moving quietly around the aisles.

The maps show the projected flight paths of enemy nuclear missiles. They track the positions of friendly and hostile shipping. There are maps indicating the precise position of enemy missile bases, troop deployments and potential lines of advance.

The room and its keepers are constantly fed with information from an array of intelligence sources, including the exotic spy satellites in space; hardware placed there at a cost of billions of dollars to keep its owners constantly appraised of the movements of their enemies and safeguard them from the threat of attack. As it arrives, information is fed into a powerful computer whose calculations allow the screens to be regularly updated. The room's keepers have the capacity to zoom in on a specific part of the globe as required. All mankind's technological sophistication is at the keepers' disposal.

Cassandra.

T.A.Große del.et sculp.

Cassandra, the Trojan princess doomed to prophesy the truth but never to be believed.

Such places do exist. NORAD, the North American Aerospace Defense Command under Cheyenne Mountain, Colorado, is one. The President of the United States has an underground command bunker in the Maryland hills and the Prime Minister of the United Kingdom has a smaller-scale version in disused limestone quarries near Bath, in Somerset. These temples to defence demonstrate how far governments are prepared to go in expending resources to provide the means to defend the nation-state: the weaponry to deter or defeat enemies, the planes and spy satellites to prevent surprise attack, to give their citizens the best chance of surviving in a hostile world.

Imagine now that just such a room exists to track environmental threats on every scale. The immediate threat to individuals from poor air quality, regional threats from the transboundary pollution of air and water-borne toxins, and global threats such as ozone depletion and climate change. It provides us with up-to-date information about what is happening to the world's environment. This threat assessment is designed to help us make decisions of a different kind about security. The same technology, the same powerful computers, remote sensing satellites, ground intelligence, this time provide data from the atmosphere, from the seas, about our forests, about air quality in our major cities, about health indicators. The maps also show strategic areas vital for the health of the planet: rainforests, wetlands, shallow ocean areas.

This is the information which governments now require in order to make decisions and allocate resources to provide for their citizens' security. These are threats without enemies. With a fine sense of irony, the designers of this "Green War Room" have called it CASSANDRA. Cassandra was the daughter of Priam, King of Troy. To win her love, the god Apollo gave her the gift of prophecy. But when she rebuffed his advances, he spoiled the gift by decreeing that she be doomed to prophesy the truth but never to be believed. She prophesied the sacking of Troy and was disbelieved. No one could imagine a deception as bold as the Trojan Horse. But she was right. Troy was sacked.

The screens of CASSANDRA

The screens on the walls of CASSANDRA show the situation for each of a number of developing environmental threats: they constantly change as environmental threats to security develop around the world. In the life span of the planet imagined as twenty four hours, we are looking at phenomena which occur in the last second of the last minute before midnight, and yet which have the power to wreck the whole ancient enterprise of life on Earth.

None of the environmental threats perceived on CASSANDRA's screens is more than two hundred years old, and many are much more recent in origin; all have begun to become serious only within the last fifty years. In 1940, the rainforests were still pretty much intact, there was no hole in the sky from ozone depletion and the idea of global warming forced by human activity was barely considered. The threats to environmental security have grown during the same life span as the atomic age but large numbers of people have only begun to

---

To read about threats to environmental security, we recommend three authoritative and easily available books:
(1) Norman Myers (ed), *The Gaia Atlas of Planet Management*, Pan Books, 1985
(2) Geoffrey Lean, Don Hinrichsen and Adam Markham, *WWF Atlas of the Environment*, Arrow Books, 1990
(3) Edward Goldsmith, Nicholas Hildyard, Peter Bunyard and Patrick McCully, *5000 Days to Save the Planet*, Hamlyn, 1990.

think seriously about them very recently, whereas we knew that we had to wrestle with the problem of the Bomb from the day after Hiroshima.

Many of the principal threats are shown here.

> ★★★★★
> **HUMAN POPULATION GROWTH**
> ★★★★★
> **MAN-MADE POISONS**
> **URBAN AIR AND WATER POLLUTION**
> **ACID RAIN**
> **THREATS TO THE WORLD'S FRESHWATER**
> **DESERTIFICATION AND SOIL EROSION**
> **POLLUTION AND EXPLOITATION OF THE ICE ZONES**
> ★★★★★
> **DEPLETION OF THE STRATOSPHERIC OZONE LAYER**
> **LOSS OF TROPICAL FORESTS**
> **LOSS OF TEMPERATE FORESTS**
> **LOSS OF WETLANDS**
> **DESTRUCTION OF MANGROVE SWAMPS AND CORAL REEFS**
> **LOSS OF THE DIVERSITY OF LIFEFORMS**
> **LOSS OF GENETIC POTENTIAL**
> **ALTERATION OF OCEAN CURRENT SYSTEMS**
> **ATTACKS LOW ON THE OCEAN FOOD CHAIN**
> **GLOBAL WARMING AND CLIMATE CHANGE**
> ★★★★★

On this visit, we shall look at just three of these screens. The screens are grouped to show the potentially different nature of the varied threats. The rise in the human population is a driving force behind the increase in environmental stress. That statement mustn't be misunderstood. It is *not* the same thing as saying that if the poor didn't breed so fast, everything would be all right. The causes of population growth are much more complicated than that. So the population screen stands separate, and it is the first one that we must look at carefully.

A large group of activities, recorded on the second group of screens, creates *pollution*. Pollution may be unpleasant, but it isn't new in the long history of the earth. All life activities, including you living and breathing as you read this page, create pollution in its most basic form, interpreted as giving out heat as a contribution to entropy (the averaging out of heat and cold). The history of the earth to be read in the pages of its geology reveals a record of amazing toughness and flexibility in its life systems, able to accommodate huge injections of pollutants and still survive, although often much changed.

It is increasingly thought that a gigantic injection of dust into the atmosphere, perhaps from a meteor strike, created a reflective blanket that kept out the sun's rays, cooled the surface and caused the extinction of the dinosaurs. But other life forms, especially warm-blooded, fat or furry mammals, rose to take over *Tyrannosaurus rex*'s crown. We have only one incident in recent recorded history which gives us any inkling of what that might have been like. This was the explosion of the volcano Tambora in 1815, an order of magnitude larger than the well-known explosion of Krakatoa in 1883 – which made 1815 famous not only for the Battle of Waterloo, but as "the year without a summer", when it snowed in Switzerland in May. Staying by Lake Geneva that year, Byron wrote his chilling poem, "Darkness". It isn't unreasonable to link the two events and to find in it the most graphic imaginative depiction that we are ever likely to have of an artificially created dark, and to see in it a sign of what the earth has been capable of recovering from during its lifetime.

> I had a dream which was not all a dream.
> The bright sun was extinguish'd, and the stars
> Did wander darkling in the eternal space,
> Rayless, and pathless, and the icy earth
> Swung blind and blackening in the moonless air.
> Morn came, and went – and came, and brought no day.

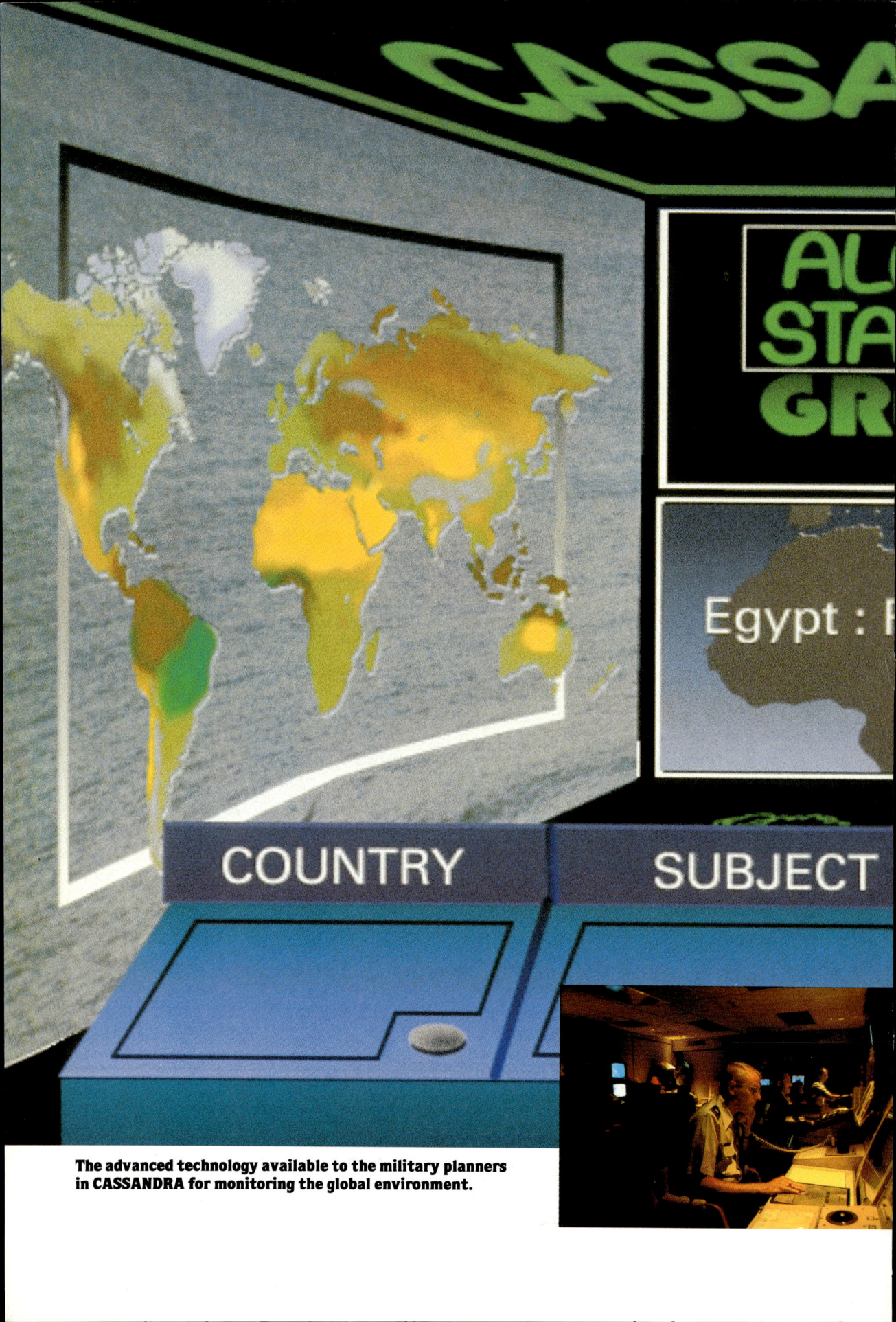

The advanced technology available to the military planners in CASSANDRA for monitoring the global environment.

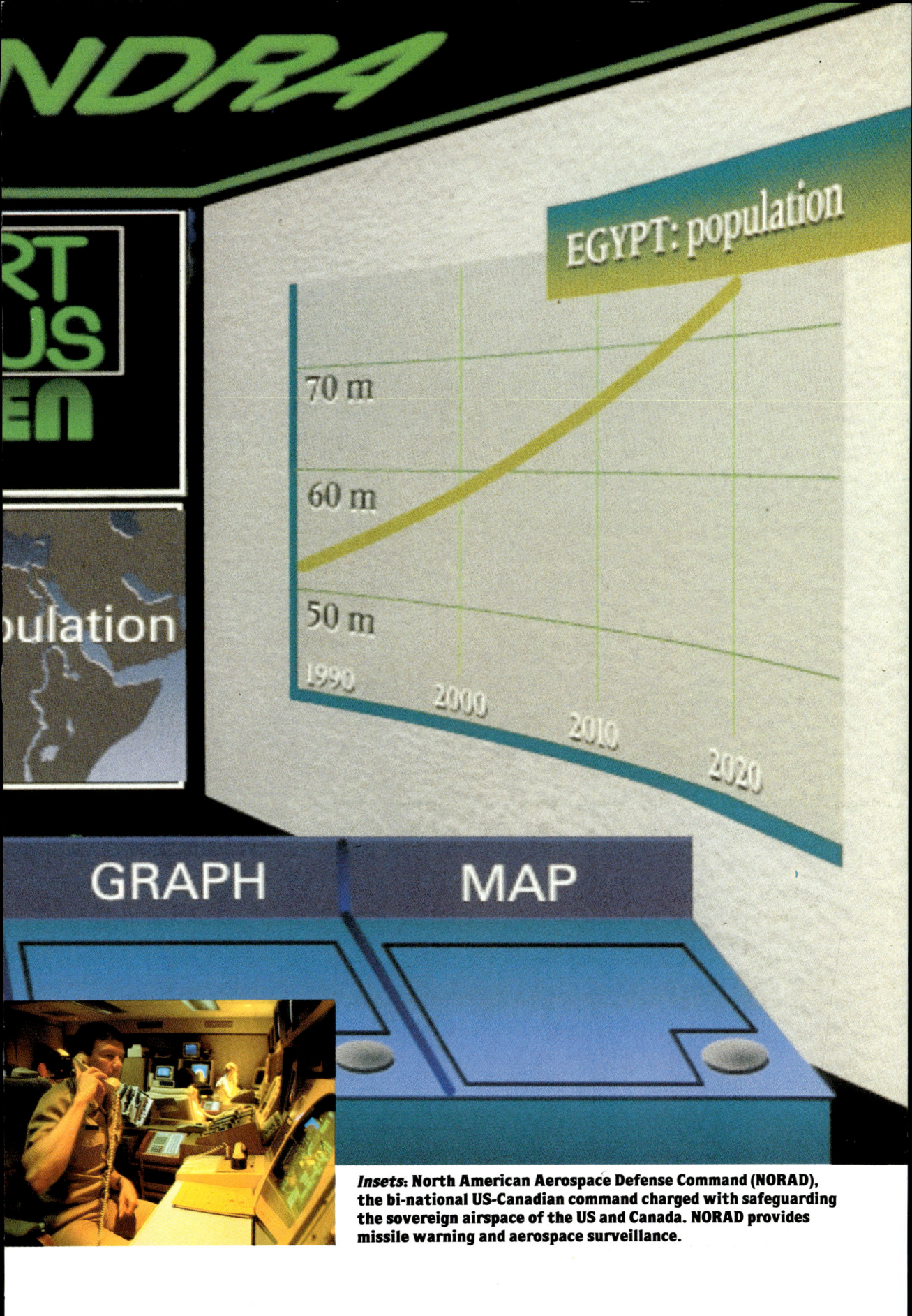

Insets: **North American Aerospace Defense Command (NORAD), the bi-national US-Canadian command charged with safeguarding the sovereign airspace of the US and Canada. NORAD provides missile warning and aerospace surveillance.**

And men forgot their passions in the dread
Of this their desolation . . .

Staying in the same house, Mary Shelley wrote *Frankenstein*.

In comparison, many of the gross pollutants (smoke, heat, dust, excess volumes of common, naturally occurring chemicals and gases) that mankind has thrown into the biosphere in the last couple of centuries may not yet be exceptional by the standards of the earth's experience. It is instructive to recall that the amount of CO_2 injected into the atmosphere from forest clearing and burning was only surpassed by that from industrial activity in the 1960s.

From all the possible pollution problems, we shall choose man-made poisons, considering chemicals and, briefly, the emerging threat of Genetically Manipulated Organisms (GMOs). The logic is that our manufacture and use of poisons is a central expression of the contemptuousness which enraged Rachel Carson: because awareness of this problem was where the modern movement of environmental consciousness began in the early 1960s; because as we shall see in Chapter 4, they can produce misery and illness if we do not act now and finally, because from among these chemicals has emerged one which has already triggered a new sort of problem altogether.

A third group of activities, portrayed on the last set of screens, may be new in Earth's experience – certainly in the speed and simultaneity of their occurrence. These are actions which may well threaten the *control systems of life* on the planet because they can interfere with the self-regulatory ways in which the balance of the atmosphere is maintained. One of James Lovelock's greatest services has been to direct our attention to this frightening possibility.

It seems increasingly that certain areas of the planet are more essential to these control processes than others, and certain man-made chemicals may be harmful out of all proportion to the volume released. These screens therefore show the highest priority threats. They are, of course, interconnected with the other screens because the polluting activities may contribute to this end: the computers that run the different CASSANDRA screens are all interactive. But for clarity, we follow Lovelock and distinguish between *pollution* and *planet modification*. It's important to know where to focus one's efforts. We shall look at the story of CFCs and the ozone hole, because it is the first of the planet modifying crises to have seized public awareness on any scale, and is unusually clear in cause and effect. It may, indeed, be the first and only warning we shall receive on how to handle the much larger and more complex issue of global warming and climate change.

HUMAN POPULATION GROWTH

The central screen in the room is the Population Screen. In the bottom right hand corner there is a counter that turns over three times a second, 180 times a minute, 259,200 each day, 94,600,000 every year. It is counting each new baby. To the left of the current population map is one with graphs on it for each area of the world showing historical population trends.

The summary graph shows that in 1950 the population of the world was 2.5 billion. In 1987 it passed 5 billion. It doesn't need a computer to show that by this doubling, the world's population has increased by the same amount as the *total increase in population from the time man first walked the earth up to 1950*. The rate and scale of these changes are unprecedented. The "take-off" moment came about 1960, when, partly under the influence of the success of "death control" by modern medicine, many more children survived to be able to reproduce than before, and more people lived longer.

But the regional graphs show that this statement has very different regional meanings. It shows that in the rich countries and in Communist countries, the birth rate fell. Different colours show different levels of population density around the world. Some evidence supports the speculation that in rich countries it fell because raising children costs so much and interferes with freer, more self-centred lifestyles, and because, after the Pill, women could control their fertility for the first time that children went out of fashion; whereas in the Communist world, women went on "birth strike" if they could, voting with their wombs, which often meant undergoing multiple abortions during their reproductive careers. When rich-world people do have children, their motives are overwhelmingly non-material. Attitudes to children in the poor world stand in sharp contrast. The Philippines case study illustrated here is fairly typical. In contrast to "birth strike" in the North, in the poor world of the tropics, birth rates exploded. So it has been there that most of the children have been born into poverty. This also produces vastly different age profiles for rich and poor.

To the right of the current population map is a screen showing estimated population trends into the next century. The overall increase anticipated over the next 35 years is from the current 5 billion to 8.5 billion. More calculations indicate that of the projected increase of 3.2 billion (which many consider to be a conservative estimate), 95 per cent of the increase will be in the developing world, where coping with burgeoning populations and resource scarcity or depletion is already a huge problem. The age-profile differences of rich and poor are set to diverge further.

The links between the demands of growing populations in the developing world and resource depletion are clear. In Lesotho, which has the highest rate of soil erosion anywhere, the demands of subsistence agriculture upon its mountainous land are to blame. In Indonesia, the felling of hardwood forests in the search for foreign exchange earnings also leads to soil loss from unrestrained run-off, but for a different reason. A high birth rate in the Indian state of Rajathstan has increased pressure on soils that are already arid. The desire of the Brazilian government to clear forest for land-hungry poor farmers, as well as to make grazing for rich ranchers' scrub cattle destined for fast-food hamburgers in North America, has been a major source of added stress on the tropical rainforests.

The links between population rise and resource depletion in the rich world are not so obvious but just as punishing. People in the richer countries often say that "the population problem" is the underlying problem that the world community must address if is to resolve any of its major international environmental issues. What they mean by that is the problem of other peoples' babies, especially black, brown and yellow babies. But it is not a question of numbers alone. Population is a complex issue. Those who worry about the number of mouths to feed in developing countries should consider the different patterns of consumption which characterize the richer and poorer countries.

Most of the natural resources of the world are consumed disproportionately by a minority of the world's population. A child born in America or Western Europe will consume far more of the earth's resources over his or her lifetime than a child born in a poor country. The Netherlands has a population density of 1,031 per square mile. This is only possible because that country imports huge quantities of essentials from elsewhere: 4 million tonnes of cereals, for example. It has been calculated that to support its population, the United Kingdom farms two hectares of "ghost land" overseas for every hectare farmed at home. So in a direct sense, each rich world baby takes up space in other parts of the world as well.

In the United States, there is an automobile for each 2.2 people. In 1982, the vehicles used 4 million barrels of oil a day. In Asia there is an automobile for each 246.7 people. These vehicles consume 0.26 million barrels of oil per day. Indeed, all the vehicles in Latin America, Africa and Asia combined use only a quarter of the energy used by the United States car fleet, per day.

By devouring energy and raw materials through our profligate life style, each new baby in the developed world weighs much more heavily upon the wasting assets of the planet than each poor-world baby. Less damaging to the general environment, each new baby in the poor world by its birth slashes at the forests and erodes the soils of its immediate environment, reducing its own chance of survival.

> "While overpopulation in the poor nations tends to keep them poverty-stricken, overpopulation in rich nations tends to undermine the life support capacity of the entire planet." (Paul Ehrlich, Professor of Population Studies, Stanford University)

It has been a general experience, especially in the rich world, that falling death rates for children under five have accompanied rising income levels and better education and that falling birth rates follow an improvement in the perinatal mortality rate. As we noted above, the liberation of women from drudgery and uncontrolled risk of pregnancy has brought about negative birth rates in some countries such as Sweden and West Germany.

In contrast, the problem for many developing countries is that they chase a rainbow. For their populations to stabilize they have to increase the confidence of each man and woman that a child born will survive to maturity, to support the parent in old age. But population growth in some areas, especially in sub-Saharan Africa, is now so great that the economies have already taxed and over-taxed their natural resource base to support the people now living, mortgaging the future. In the Sahel, in the valley of the Ganges in India, on the island of Java, Indonesia, or in the Andean highlands of South America, the increase of poverty, population and environmental degradation are viciously intertwined.

Half the world's poor live in South Asia, home to 30 per cent of the world's population. A further quarter are in East Asia, where 40 per cent of the world's population live. For Africa, the figures are 16 per cent and 11 per cent respectively. The burden of poverty is carried disproportionately by women and children. And here is the key.

It is in the liberation of the *minds* of women that the most reliable means exist to help them to control their bodies and their lives and through so doing, to reduce population growth. A focus on contraception as a technical problem in isolation from everything else simply doesn't work.

The South Indian state of Kerala has three times the population density of the all-India average, yet one third the infant mortality rate. Why? To be sure, Kerala has had for forty years very active grass-roots politics, which have led to good rural health care, to co-operatives, Credit Unions and to land reform. But crucial for its success in population management, it has for generations in Kerala (predominantly a Christian area) been thought to be right to educate girls as well as boys. So there is a very high female literacy rate, two and a half times the all-India average. It correlates with the lowest perinatal mortality rates, the lowest birth rates, the longest life expectancy in the region and also, unquantifiably, with

Visiting CASSANDRA

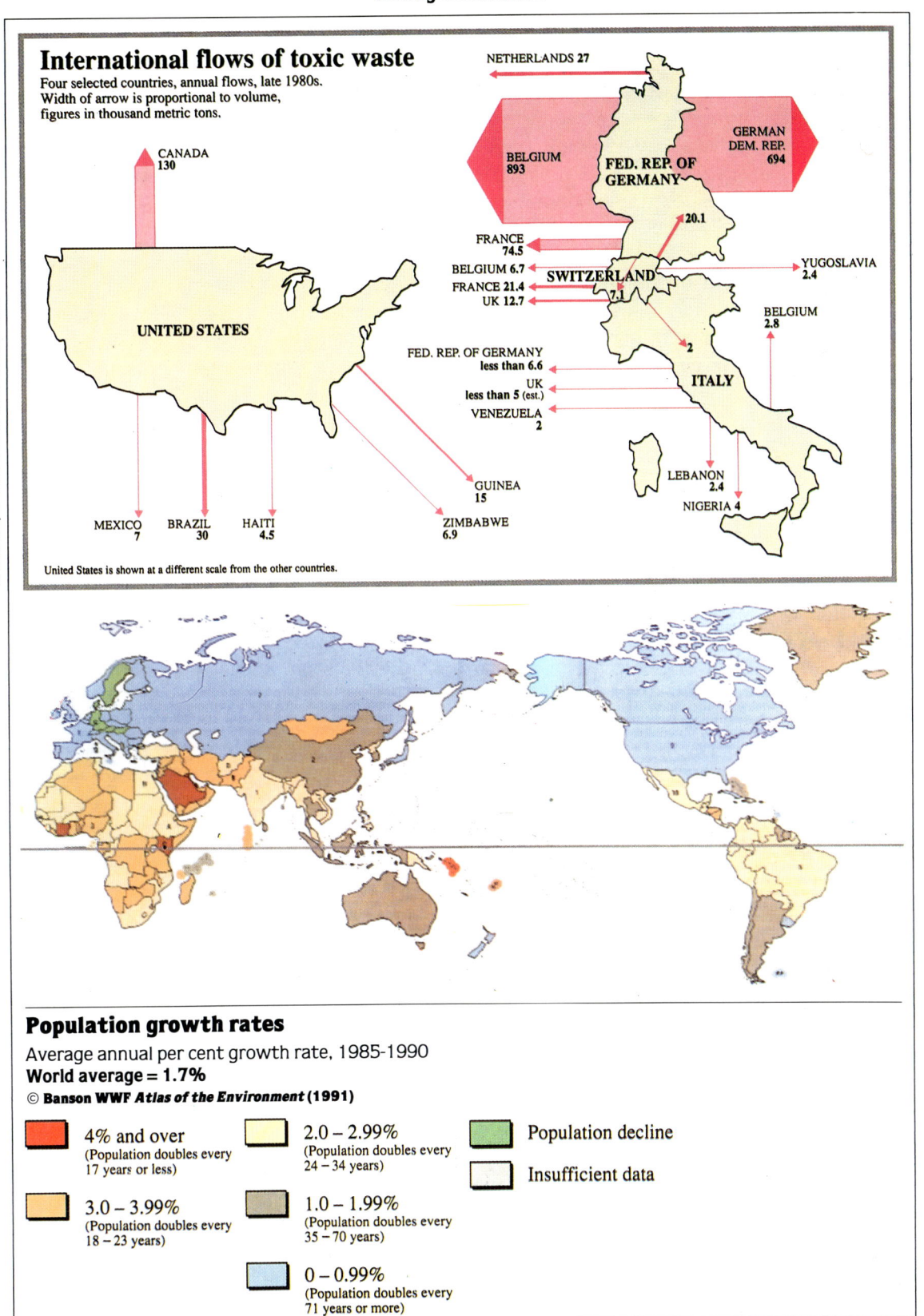

International flows of toxic waste

Four selected countries, annual flows, late 1980s. Width of arrow is proportional to volume, figures in thousand metric tons.

- NETHERLANDS 27
- BELGIUM 893 → FED. REP. OF GERMANY
- GERMAN DEM. REP. 694 → FED. REP. OF GERMANY
- FED. REP. OF GERMANY → SWITZERLAND 20.1
- FRANCE 74.5
- BELGIUM 6.7
- FRANCE 21.4
- UK 12.7
- YUGOSLAVIA 2.4
- SWITZERLAND 7.1
- BELGIUM 2.8
- ITALY 2
- FED. REP. OF GERMANY less than 6.6
- UK less than 5 (est.)
- VENEZUELA 2
- LEBANON 2.4
- NIGERIA 4

UNITED STATES:
- CANADA 130
- MEXICO 7
- BRAZIL 30
- HAITI 4.5
- GUINEA 15
- ZIMBABWE 6.9

United States is shown at a different scale from the other countries.

© Banson, WWF Atlas of the Environment

Population growth rates

Average annual per cent growth rate, 1985-1990
World average = 1.7%

© Banson WWF *Atlas of the Environment* (1991)

- 4% and over (Population doubles every 17 years or less)
- 3.0 – 3.99% (Population doubles every 18 – 23 years)
- 2.0 – 2.99% (Population doubles every 24 – 34 years)
- 1.0 – 1.99% (Population doubles every 35 – 70 years)
- 0 – 0.99% (Population doubles every 71 years or more)
- Population decline
- Insufficient data

a better, more fulfilled life for men and women both. None of this was (or could be) done by programmes of forcible sterilization or the promotion of contraception in the absence of a total development package. Female literacy rates are a powerful indicator of the wellbeing of a population. So another of CASSANDRA's screens carries such information.

Under the discipline of environmental security, the question which population pressure poses to each country in the world is the same. How to attain a sufficient sense of security and wellbeing for each individual to wish to have a smaller family, without destroying the environment in the process. This is as true for the USA where gross national product per capita is $19,840, as it is for Kenya where it is $370 or Mozambique where it is $100. We cannot aim for the American level of consumption as a world-wide goal. The biosphere couldn't stand it; nor can we leave the poor to languish in the trap. To do either is to invite the Horsemen of the Apocalypse to solve the problem for us.

In his *Second Essay on the Principle of Population* of 1803, the Reverend Thomas Malthus offered mankind a stark choice. Malthus suggested a general principle: that "population, when unchecked, increases in a geometric ratio. Subsistence only increases in an arithmetic ration." Malthus thought that the power of multiplication could be constrained by only three things: checks of vice and misery (War, Pestilence and Famine) or checks of moral restraint. Malthus had in mind sexual abstinence. Environmental security concerns enlarge our definition of public morality to embrace the health of the planet, and the challenge is to make this global morality effective in our most intimate concerns.

MAN-MADE POISONS

The skull and crossbones on a rusting barrel of toxic waste has become a symbol of man's pollution of the planet. CASSANDRA's Poison Screens show that all over the world, hazardous waste dumps and the careless or inefficient disposal of chemicals are poisoning groundwater, fouling the air and causing illness among the animal and human populations. No industrialized country is guiltless.

Mankind makes regular use of 60,000 chemical compounds for everything from fertilizer to refrigerant, from hamburger packaging to pesticides. Most man-made chemicals are harmful in some degree to humans, plants and animals. Most of these substances eventually find their way into our soil, air and water. But still, thirty years after Rachel Carson first sounded the alarm in *Silent Spring*, astonishingly few chemicals have been tested for their potential carcinogenic and toxic effects. According to the *WWF Atlas of the Environment*, 100 chemicals were selected at random by the US Academy of Sciences. Extensive computer literature surveys were conducted to find out what epidemiological or toxicological data was available. Only 10 per cent of all the pesticides investigated, 2 per cent of cosmetic ingredients, 18 per cent of drugs and 5 per cent of food additives had sufficient data on them to enable complete health assessments to be made.

What we do know is that many synthetic organic chemicals including the most dangerous compounds, such as the pesticide DDT or the fire retardant Polychlorinated Biphenyl (PCB), build up in the fatty tissue of animals. The concentration of pollutants in animals increases as it ascends the food chain, from predator to predator, as we saw with the toxic whales of the St Lawrence river in Canada. Seals in the North Sea sit at the top of their food chain, just as human beings do theirs. The seals have shown concentrations of PCBs millions of times higher than are found in the seawater around them. They have also become less able to breed.

The United States alone generates roughly 240 million metric tons of hazardous waste a year. The Organisation for Economic Cooperation and Development estimates that 330 million metric tons are produced world-wide. A study by Friends of the Earth and the *Observer* newspaper identified 4,800 toxic tips in the UK. But it is the USA that of all the countries in the world has most to 1ear from decades of careless dumping of waste in landfills, ponds, fields, rivers seas and lakes. The US Office for Technology Assessment estimates that there could be 100,000 sites which require immediate attention. It would take 50 years and $100 billion to clean them up.

As more information has become available it is clear that Eastern Europe, particularly the "black triangle" composed of the former DDR, Poland and Czechoslovakia, is a toxic horror story. Waste sites such as the one outside Usti nad Labem in northern Czechoslovakia are hideous scars on the landscape and a standing reproach to Europe. Huge numbers of rotting, leaking barrels of toxic waste litter the steep sides of a deep pit into which toxic waste has been dumped over the decades with no attempts at control whatsoever. The pink and purple and white mixture that lies at the bottom of the pit and leaches into the water supply contains a chemical cocktail of unknown composition.

As public pressure and regulation make irresponsible dumping harder and harder in industrialized countries, a particularly nasty aspect of toxic waste dumping has developed. The rich and powerful have found that selling their waste to the poor is good business. So in the famous case of the ship, the *Karin-B*, toxic waste of Italian origin turned up in West Africa and, when discovered, had to be removed back to Europe in the *Karin-B*, which could find no port to accept her.

The United States exports a total of 63,000 metric tons of waste to countries including Haiti, Zimbabwe and Guinea. None of these countries have the technology to deal safely with dangerous chemicals. Italy exports the bulk of its toxic waste to Venezuela, the Lebanon and, until the *Karin-B* episode, Nigeria. The developing world has been used as a convenient waste tip for toxic materials and the Bhopal accident which is described in Chapter 4 was a dreadful example both of the failure of a multinational company to tell the host country exactly what was going on in the plant, and of failed procedures in the plant. Worse may yet be coming, as scientists develop their ability to manipulate both man-made and natural substances.

Plant breeding holds the promise of great benefit for the poor world, but also great danger. The most famous example so far has been Dr Norman Borlaug's work on super strains of rice, which gave India the "Green Revolution" but also immense social dislocation and increased dependence on high input farming techniques. Science has moved forward rapidly, and the frontier stands with biotechnology: actively manipulating the genetic coding of living organisms.

Abigail Munson, a Cambridge researcher, is looking at the possibility that Genetically Manipulated Organisms – products of this current revolution in biotechnology – may follow the career of chemicals in the Third World. GMOs are plant and animal substances whose genetic composition has been tailored to favour a particular characteristic. This might be drought resistance in a cultivar like wheat, for example. Or increased lactation in a cow injected with synthetic BST (Bovine Somatotropin), a growth hormone. The Response Strategies Working Group of the Intergovernmental Panel on Climate Change in 1990 viewed the possibility of GMOs assisting in crop adaptation to drought with encouragement. What Ms Munson is finding, however, is that for a science with such amazing and fundamental powers to change, to improve or to blight life, there is as yet almost no regulatory framework which can protect the poor world from decisions to release GMOs, by accident or by design. The CASSANDRA poison screens therefore carry a sober warning for the management of GMOs, still in its infancy.

THE HOLE IN THE SKY

One of the maps monitored most carefully in CASSANDRA shows levels of stratospheric ozone, which have been watched closely ever since 1984. While making routine measurements at their base at Halley Bay, Dr Joe Farman and the team from the British Antarctic Survey discovered that the ozone which should have been above them had literally vanished. The NASA computers, which originally appeared to contradict the findings, later supported them. The reason why NASA missed the significance of the data in the first place is in itself instructive. NASA's Nimbus 7 satellite had also detected the ozone depletion; but the computers into which its data was fed had been programmed to reject any "aberrant" information which seemed to deviate too widely from reasonable expectation.

Since the early 1970s some scientists had warned that the ozone layer could be gradually eroded by chlorofluorocarbons (CFCs) and other gases but very few had predicted anything on this scale. In 1989 in addition to the Antarctic ozone hole which was bigger than many had predicted, an alarming discovery was made of a 30 per cent loss in the latitudes between 50° south and 60° south, areas well outside the polar region where the particularly cold conditions precipitate the breakdown of compounds containing chlorine.

CFCs were a source of great pride for their inventor Thomas Midgley, back in the 1930s (Midgley was also the inventor of leaded petrol). They are inert gases, very "safe" to handle, and so were used as aerosol propellants, in making blown foam packaging, in coolants for refrigeration and air conditioning, and in ultra-clean industrial processes, such as in the production of microchips for computers. It was only after decades of their use that their deadly chemical properties in places far away from where their inventors used them – literally over their inventors' heads – were fully understood.

When it was first suggested seriously that using an underarm deodorant might actually help to threaten life on earth, the idea seemed to be too preposterous to "realists". If *that* were true of something as harmless and civilized as personal hygiene, what would those wretched doomsters, so intent on spoiling our innocent pleasures, complain about next? Don't they think that anything is safe? The general reaction was a loud, if nervous, guffaw. But it was true.

Certain CFCs have a very long life in the troposphere and drift up into the stratosphere. As the ultraviolet light begins to break them down they release chlorine atoms. This reacts with the natural ozone layer, destroying it and forming in its turn oxygen and chlorine monoxide. The chlorine monoxide then reacts with an oxygen atom forming an oxygen and a chlorine atom. The chlorine is not used up in the cycle and is thus capable of going on destroying ozone over and over again.

CFCs are not the only gases involved in the destruction of ozone. Halons used in fire-fighting equipment, methyl chloroform used in cleaning solvents, adhesives and aerosols and carbon tetrachloride used, among other things, for dry cleaning, all release chlorine or the related gases bromine and fluorons. Nitrous oxide is another culprit.

NIMBUS-7: TOMS TOTAL OZONE

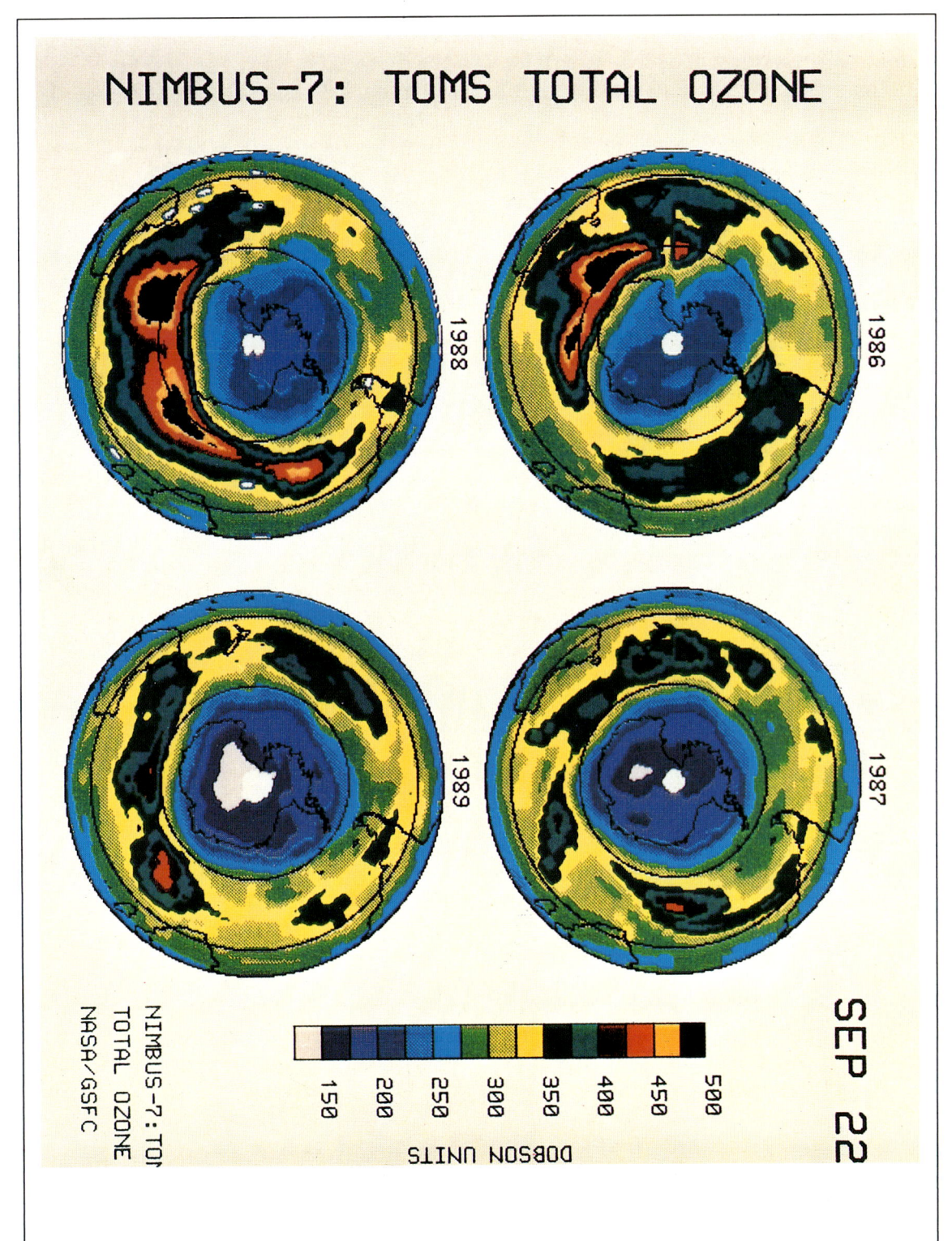

SEP 22

The ozone hole over the Antarctic: the first major "Show Me" Crisis evidence of the assault by man-made chemicals on the atmosphere.
© NASA

Ozone plays a vital role in protecting life on earth. It is a form of oxygen with three atoms instead of the normal two. It lies in a thin skin 15-50 kilometres up in the stratosphere. Ozone forms a shield from the lethal short wave Ultra Violet-B solar radiation rays of the sun. Without this protection, UV-B radiation would wreak havoc. We know enough about the damage that the small amount of UV radiation which does reach the surface causes that we should be deeply concerned about potentially significant increases.

UV-B is an important cause of skin cancer. The American Environmental Protection Agency calculates that for every one per cent decrease in the concentration of ozone in the stratosphere there will be a 5 per cent increase in the number of non-malignant skin cancers each year in the USA. Further studies suggest that malignant melanomas would increase by one per cent, causing as many as 12,000 extra deaths in the USA alone. In fact, with the erosion of the southern ozone shield of the last decade, blue-eyed, blonde and red-headed white skinned people in Australia, New Zealand, South Africa and South America who do not protect their skins from direct sunlight are most at risk.

Plants too are susceptible. Protein-rich soy beans for example, one of the great successes of modern agriculture, suffer a 25 per cent decrease in yield from a 25 per cent increase in exposure to UV. Life in the oceans is vulnerable as well. Phytoplankton, the tiny organisms at the bottom of the ocean food chain – that is, upon whom the whole ocean life pyramid stands – suffer particularly badly. The consequences of this may be immense. Not only would a decrease in phytoplankton work its way up the food chain and onwards to the human beings who depend upon food from the seas, but if Lovelock is correct in hypothesizing a role for them in climate control, there might be an unforeseen, synergistic effect upon the balance in the composition of the atmosphere.

Of course, we can't be absolutely sure of all these predictions. This is a lot to do with the nature of injuries to the natural world. They do not mount in severity at a steady pace. It is not the case that the accumulation of poisons in the soil or of greenhouse gases in the atmosphere produce a congruent increase in effects. The characteristic shape of the line showing cause and effects is not the arithmetic straight line but instead, the far more insidious and dangerous *exponential curve*.

The problem about exponential curves was made clear by pioneering scientists looking at the interactive effects of resources, population and pollution on the planetary system in the early 1970s. The path-breaking study *Limits to Growth*, the 1972 report to the Club of Rome by Professor D. Meadows and others, shows that for a long period of time, as a pollutant or other causes of stress accumulate, there is no visible change in circumstance. But there comes a moment – a moment which is not predictable before it occurs – when suddenly the system accelerates away from the earlier state of what had appeared, misleadingly, to be stability.

What James Lovelock was pointing out with his Gaia hypothesis was that the ability of the biosphere to tolerate stress and damage and changing circumstances, whether from cosmic or from human cause, is quite immense. His particular hypothesis is that one reason for the vast range of tolerance within the biosphere is that it acts as if it were an organism itself, and therefore in many and complicated and beautiful ways "feedback" mechanisms are brought into play to restore stability after a change in input. Phytoplankton are one of the agents which he identifies as being potentially very important. But, Lovelock warns, in common with the scientists of the *Limits to Growth* study twenty years before, the moment may come when the system is finally overloaded and the feedback control mechanisms simply cannot cope. That is the moment when the exponential curve goes "vertical"; that is the moment when the control mechanisms all of a sudden can no longer control. That is why there will continue to be scientific doubt arising from a lack of definitive observables for a considerable time in many areas of environmental stress. But as the Prince of Wales, known for his concern over environmental issues, has observed: "If science has taught us anything it is that the environment is full of uncertainties . . . while we wait for the doctor's diagnosis the patient may easily die". To receive any warning at all that something is going wrong in the atmosphere is a bonus.

So the hole in the sky matters for another reason too: a political reason. Steven Schneider, Head of the Interdisciplinary Climate Systems Programme at the National Center for Atmospheric Research in Boulder, Colorado has been prominent in the attempt by climate scientists to alert the public to what they see to be the present and real threat to the atmosphere. But the problem, as Schneider points out, is one of helping people to make acts of considerable imagination, exemplified by the licence plates of cars in Missouri. Missouri is the "Show me!" state. People demand a "Show me!" crisis, Schneider observes, and the problem about climatic change is that we cannot yet really offer conclusive evidence of "Show me!" crises. We can't say, "Look! The sky has turned yellow!" or "See! You simply can't breathe!"

The ozone hole over the Antarctic, however, has given us one graphic "Show me!" crisis. It may be the only "Show me!" crisis that we shall have and we would be wise to pay attention to it.

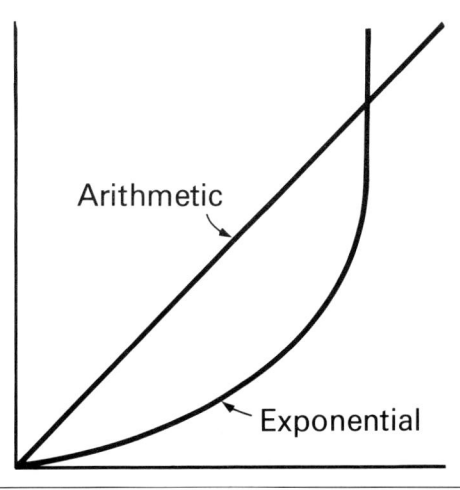

Toxic waste dump, Koko, Nigeria.

At the bottom of the food chain, phytoplankton (*opposite*) are also an important indicator of planetary health.

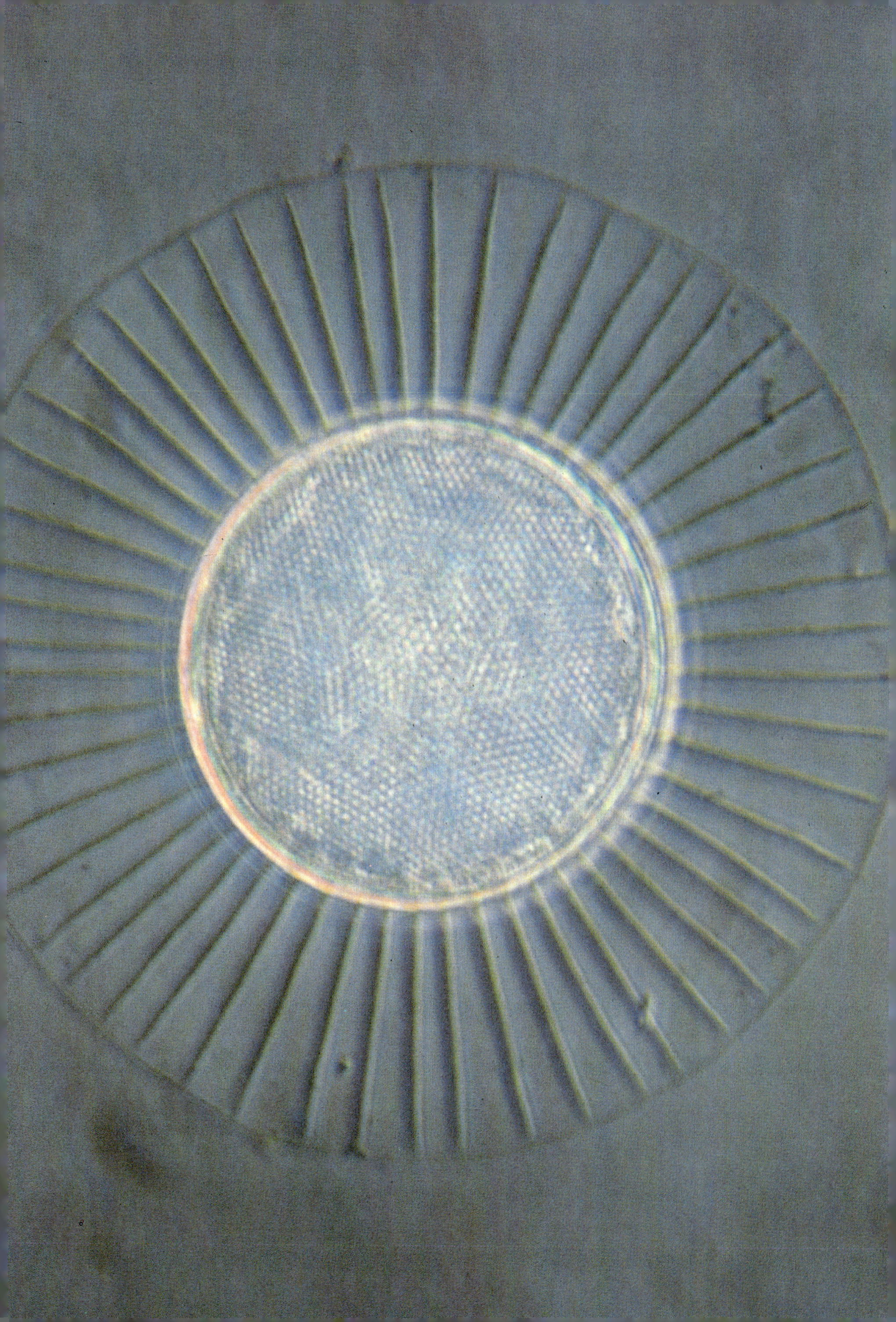

Chapter 3: "De Nile Ain't Jus'a River in Egypt"

Four reasons for disbelieving CASSANDRA

So what do we make of these threats without enemies that we have seen up on the screens in CASSANDRA?

Lester Brown, head of a leading environmental organization and one of America's most outspoken environmentalists, echoes the feelings of many of the specialists:

> Each year at the World Watch Institute in Washington we in effect give the earth an annual physical examination. We check its vital signs and each year the patient's health deteriorates further. The forests shrink, the deserts expand, the topsoil erodes, the ozone layer is being depleted, the number of plant and animal species with which we share the planet is diminishing. We now have acid rain on every continent. Air pollution has reached health threatening levels in hundreds of cities . . . each year things grow worse. At some stage in the not too distant future we have to turn these things around. It is time to put the patient into the intensive care unit.

But up to now the urgency of this rather grim diagnosis has, frankly, not been widely shared. Why is this?

When Lieutenant James Cook entered Botany Bay in Australia in 1770 on board HMS *Endeavour* on the first of his epic voyages of exploration to the Pacific, the aboriginals on the shore did not even look up as the great ship entered the bay. Joseph Banks, Cook's companion, wrote that, "the ship passed within a quarter mile of them and yet they scarce lifted their eyes from their employment . . . expressed neither surprise nor concern." Yet when boats were lowered to row ashore, then the aboriginals sprang to defend themselves. What does it mean? It means that a ship the size of the *Endeavour*

The Shadouf has been used to draw water from the Nile for thousands of years.

Captain Cook increased our knowledge of the Pacific World, but in the end it killed him. As latter-day Captain Cooks we run the risk that the same fate could overtake humankind on a global scale. The more we learn, the sooner we act, the smaller the risk.

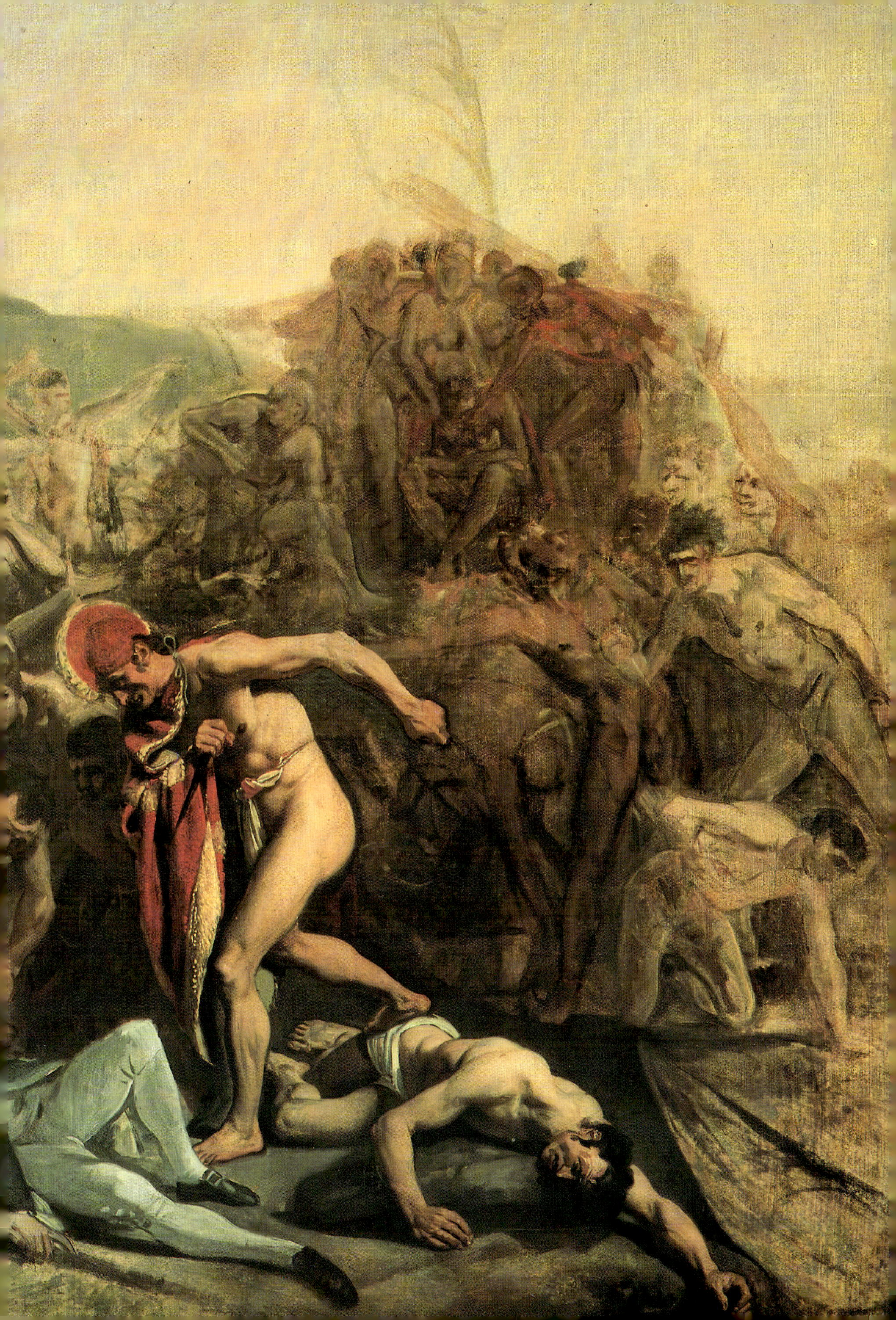

was so far beyond the ken of the aboriginals that they simply couldn't imagine the possibility of such a thing and so they didn't "see" it. Of course, they saw it in a physical sense but they didn't see it in a sense which provoked reaction. However, when the much smaller rowing boats, close to the size of their canoes, were lowered into the water, they were able to see a threat which fitted more closely the scale of their lives and their expectations, respond to it and leap to their own defence.

Standing in CASSANDRA, it isn't too surprising if many of us find that the mind-boggling complexity of the many threats which the screens throw up are simply beyond what we are used to imagining, and so, like the aboriginals in the eighteenth century, we cannot "see" what is before our eyes. Lester Brown again:

> As we look at the complexity of the environmental threats and the difficulty of mobilizing the world to respond . . . we have to ask ourselves whether evolution has prepared us well for this role. We evolved over a few million years in small groups, tribal societies . . . and our behaviour, our reflexes, our instincts were shaped by the needs of survival in that environment. Now suddenly we are faced with the need to mobilize the world to stabilize climate, to stabilise population. We begin to wonder whether in an evolutionary sense we are up to it.

If the threat has an enemy behind it, then we know how to act. Aggressors are to be deterred or if they cannot be deterred are to be repulsed. But here we are being asked to make causal links which say that with the increase of world population and with the fact that each person in the wealthy world places a disproportionately greater burden on the biosphere, almost every action of "normal daily life" is suddenly in question. Its long-term effects, often through complicated chains of consequence, must all be considered. In CASSANDRA, we are in effect being asked to make radical readjustments in many of the basic assumptions with which we organize our world.

What is a private act? What is a public act? It seems that environmental security reduces greatly the realm of harmless private acts – the basis of liberal philosophy ("you can do what you like so long as it doesn't involve standing on other people's toes") – because everything that we do seems attached to and affects everything else. If that is so, what then of our old political distinction between "left" and "right"? None of that seems likely to work very well either. These problems call simultaneously for extensive and public planning of the allocation of goods and services and resources on the large scale – traditional "left wing" themes – and for vigorous and energetic entrepreneurial innovation and private initiative where the individual takes charge of decisions about an even wider range of actions in his or her life than politics has traditionally regarded as being necessary – traditional "right wing" themes. So for modern aboriginals like us, we can see how it has been difficult to imagine the environmental security problem.

There is a second reason why people may not share Lester Brown's view in large numbers. It is that they simply *don't see any evidence in the world around them,* as they go about their daily lives, to support the apocalyptic interpretation that is being made. What's all this stuff about putting the "patient" into "intensive care"? The skies and waters are still blue, the fields are still green. Massive environmental damage such as has occurred in the "black triangle" of central Europe remains the exception and not the rule on the face of the planet.

After the Chernobyl nuclear accident in

1986, the evacuation of villages and farms in the area of highest fallout from the exploded reactor was ordered. The bewildered peasants who lived in these places were swept out of their homes by chemical troops of the Soviet armed forces dressed in protective gear. They were taken many miles away to new homes and their old homes were boarded up. It was a sign of *glasnost* that a Russian documentary film maker was allowed access to the contaminated zone. He interviewed a frustrated officer in control of troops who had been given the thankless task of preventing the peasants returning to their homes.

"What are we to do? These people do not believe that there is anything wrong with their homes or their gardens because they cannot see that there is anything wrong. They see the notices that we have put up saying that entry to this zone is forbidden and they simply cut through or climb over the fences and go back to their own lands." So indeed they did.

"What are you doing here?" asks the interviewer of an old woman who has broken through the cordon and returned to dig potatoes in her field. She has radiation sores on her hands. "Ah," she replies. "They tell us that there is something wrong with our land and that we cannot grow vegetables or live in our homes any more. But what I know is that I need to harvest my potatoes. I see nothing wrong with them and nothing wrong with my field so that is why we are here." An old fisherman asked the same question replies more angrily, "They tell me that I shall die if I eat the fish which I am catching here. But I am old, and I will die anyway. I have caught fish in this stream for all my life and I have no intention of stopping now. If I die from eating my fish then so be it."

People want what Steven Schneider called "Show me!" crises before they will act. But, as we saw while visiting CASSANDRA, they are hard to come by, and when they do appear, are usually following *after* the moment of irretrievable change.

There is a third set of reasons that people use to avoid coming to Lester Brown's conclusion when they visit CASSANDRA. These reasons follow on from the lack of a "Show me!" crisis. They stem from the fact that there is still great scientific doubt about the nature of the problems that have been displayed on the screens, especially those such as global warming which depend upon deductions from vast bodies of observation data which then have to be computed and calculated and integrated in extremely complicated ways.

During the studies made by the different working groups of the UN Intergovernmental Panel on Climate Change during 1990, it was a repeated assertion by the Saudi Arabian and American representatives in Working Group Three, which was instructed to look at policy response strategies to climate change, that the phenomenon was not yet sufficiently proved scientifically to warrant serious reaction. In contrast to the Science Working Group (Working Group One) chaired by John Houghton the Director of the UK Meteorological Office, Working Group Three never used the words "global warming" referring instead to "climate change". It called for more scientific research to be done, whereas Working Group One had made it clear that the narrowing of scientific uncertainties in the 1990s would be only a marginal gain on present knowledge.

There is no doubt that the rapid accumulation of large amounts of further and accurate data on many of the well-known observables of the climate, as well as new series of data, is a necessary security requirement, an environmental security requirement; and later in this book we will be suggesting ways in which that process can be accelerated. But to say that there is as yet no "Show me!" evidence and that there

Radioactive waste – a chronology of mismanagement

The chief characteristic of the history of radioactive waste management in the UK has been the complete failure of the Government and the industry to formulate a responsible and realistic radioactive waste management policy. The chronology of events concerning this issue repeatedly demonstrates a deep inability to take into account the concerns of the public of the technical reality of the problems created by nuclear waste. The result has been policy reversal after policy reversal and an understandable lack of confidence in the authorities' ability to handle such a serious issue.

1976
The sixth report of the Royal Commission on Environmental Pollution concludes that "it would be irresponsible and morally wrong to commit future generations to the consequences of fission power on a unless it has been demonstrated beyond reasonable doubt that at least one method exists for the safe issolation of these wastes for the indefinite future".

1978
The Government establishes the Radioactive Waste Management Advisory Committee to sponsor research, following the criticism of the Royal Commission. Test-drilling programme commences to find a deep disposal site for high level waste.

1981
Test-drilling programme abandoned after intense public opposition, especially at the public inquiry at Mullwarchar in Galloway.

1982
United Kingdom Nuclear Industry Radioactive Waste Executive (NIREX) formed to implement the Government's new policy for the disposal of low and intermediate waste following the failure of the high level waste programme.

1983
The London Dumping Convention passes moratorium on the dumping of radioactive waste at sea.

1984
ICI, the owners of the potential site at Billingham, decide not to cooperate with NIREX because of the strength of public opposition to the plan. Some 83,000 people in the Cleveland area sign a petition of protest sent to the Prime Minister.

1985
Government announces that NIREX to abandon further investigation of the Billingham site.

1986
NIREX announce three further sites in addition to Elstow for investigation for shallow burial of nuclear waste: South Killingholme in Humberside, Bradwell in Essex and Fulbeck in Lincolnshire. Protest groups form at each site, under the umbrella group Britain Opposed to Nuclear Dumping (BOND). Local authorities form coalition to oppose dumping.

1987
Immediately before the General Election, the Secretary of State for the Environment announces that all four sites are to be abandoned as potential nuclear waste dump sites. Nirex directed by the Government to find a single, national, deep disposal facility for nuclear waste.

Greenpeace, Friends of the Earth and Cumbrians Opposed to a Radioactive Environment publish *Radioactive Waste Management – The Environmental Approach*. This criticizes NIREX's deep dumping plans and puts forward the only environmentally sound method of dealing with the radioactive waste problem: that is, above-ground or near-surface passive storage at the site of production in a monitorable and retrievable form. This policy endorsed by 34 local anti-dumping groups across the country. These groups pledge their commitment to opposition to the dumping of nuclear waste, by whatever method, anywhere in the UK.

The rest of 1988 and 1989 saw an enormous growth in the strength of public opposition to NIREX's latest plans, and support for the environmental approach advocated by Greenpeace and Friends of the Earth. Local authorities all over the UK, who were consulted by NIREX, rejected their plans, with the single exception of Copeland District Council in Cumbria. Many of these authorities support the policy of on-site storage, advocated by the environmental groups.

In November 1988 the results of NIREX's consultation exercise were published, having been analysed by the University of East Anglia. The results demonstrated that NIREX's attempts to win public support for deep disposal had failed. No clear mandate for the three deep disposal options emerged.

March 1989
NIREX announces that Sellafield and Dounreay are selected.
Greenpeace and Friends of the Earth jointly publish *Exposing the Faults*, a critique of the geological basis of NIREX's plans.

July 1989
International Atomic Energy Agency (IAEA) review NIREX's proposals and conclude that the safety case is at a very early stage of development involving a lack of field data.

July 1989
Drilling of first exploratory borehole at Sellafield begins. Originally Cumbria County Council refused planning permission; this was overturned by the then Secretary of State for Environment, Nicholas Ridley.
Highland Regional Council reject planning permission for boreholes at Dounreay.

Drilling at Sellafield abandoned due to drilling difficulties.

January 1990
Planning permission granted for second borehole to east of Sellafield.

June 1990
Planning permission granted for up to three boreholes at Dounreay by Secretary of State for Scotland, Malcolm Rifkind, overturning the Highland Regional Council's decision.

August 1990
Drilling of second borehole at Sellafield begins.

October 1990
Second Sellafield borehole penetrates basement rocks. Planning permission granted for another attempt to drill at the original Sellafield site. Drilling at Dounreay begins.

November 1990
The Radioactive Waste Management Committee publishes its eleventh Annual Report. NIREX's timetable is described as "unrealistic and flawed". Target date of 2005 seems unlikely.

January 1990
Planning permission given for three more boreholes at Sellafield – to begin in April.

remains much scientific doubt is to fly in the face of what we know to be the characteristic pattern of problems in the natural world as they occur, the prevalence of exponential curves, explained in the previous chapter.

So standing in CASSANDRA, we may find ourselves like latter day aboriginals, unable to imagine the complexity of the interactions which the different screens are asking us to grasp. Or we may be unmoved by what we see because most of it is simply speculation. We see no "Show me!" crisis and our doubts may be amplified if we look into the scientific basis of what lies behind the screens. We may have doubts because we see that much of what we are told at the moment is extrapolation from data which in itself is complicated and rarely conclusive. We can't see it; we don't see any visible evidence; there is too much scientific doubt to make it worth bothering. There may be a fourth sort of reason on top of these three why we would not find the information on the screens persuasive.

This is that we, as ordinary people, are told that we do not need to worry about these things. "They" are the people who are in power. "They" have the experts. "They" have the information. We, in contrast, are only worried individuals, people who do not have expert knowledge, people who are very much aware of our powerlessness. That, after all, is what we have governments for. It's part of the social contract that we gain power by giving power to them and in issues of scientific doubt and of environmental security until very recently the consistent line has been to tell us not to worry.

The plume of radioactive fallout from the Chernobyl accident had various curious characteristics as it was blown westwards. Radioactivity fell in parts of Italy and Switzerland and it fell also, as was later discovered, with special intensity on the hills of Wales. But – strangely – radioactivity did not fall from the cloud as it passed over France or over England at that time. Why? Because officials from the British National Radiological Protection Board and representatives of the French government went on to radio and television to tell people that there really was no cause for concern; that there was no threat to public health; that everything was under control.

Of course there was fallout over France and there was fallout over the rest of the United Kingdom. But the information about where the "hotspots" of fallout occurred was kept a closely guarded secret and only came out later as a result of the work of persistent journalists and courageous scientists. It was interesting to see that in the two countries which were most strongly wedded to their civil nuclear power programmes (France and the UK) the overriding concern appeared to be to ensure that public anxiety was not unduly aroused. And the other interesting thing is to see how to a very great extent this reassurance was accepted by large numbers of people.

It now turns out that the official reassurances that were given at the time of Chernobyl were incorrect. And indeed, especially so in the Soviet Union. The degree to which the accident has left the Soviet Union with areas of irradiated rich farmland which cannot be safely worked for generations is now beginning to become clear, as is the irresponsibility of the decision taken by some official somewhere to authorize the mixing of radioactive dairy produce from the Ukraine with unpolluted sources of produce. Produce that was then distributed throughout the Soviet Union. In the case of radioactivity, the solution to pollution is not dilution.

Yet there is something odd about the bland acceptance of the reassurances at the time. When we compare this with the record of public opinion on nuclear energy we see that, offered a choice, people prefer not to have this type of technology and the strength of their feeling increases the more visibly the power station looms on their back doorstep.

In the UK case, when an attempt was made by NIREX, the body in charge of disposing of radioactive waste, to investigate the geological properties of three sites in which to store nuclear waste, local action groups sprang up in the different places under threat like dragon's teeth and successfully mobilized a national campaign which led to the abandonment of the attempt.

So the interesting thing is to see that people who were prepared to accept the general reassurances from those in power in general and abstract terms suddenly switched from trust to deep mistrust when the issue came into their own backyards. Why so?

In one sense, it is like the moment when HMS *Endeavour* lowers its rowing boats to come ashore. Suddenly the aboriginals on the beach can recognize a threat that is within their own scale and they can set about finding ways to fight it. But there is another and deeper reason which lies below all those which we have already given to explain why the information on the screens in CASSANDRA has not resulted yet in dramatic change in the way we live our lives.

The Perils of De Nile

> "My friend Carl Sagan is fond of quoting Dire Straits, the British rock group. 'De Nile ain't jus' a river in Egypt . . . denial is a psychological response whereby people pretend a problem doesn't exist . . .'" Senator Al Gore, December 1990

People everywhere have basic needs that nobody disputes. There are physical needs for food, for shelter, for warmth, for clothing. There are social needs for reproduction, for friendship, for society, for respect, for identity in a group. And there are psychological needs which are not as obvious as the others but which are quite as important. Pre-eminent is the need to feel that one is in control of one's situation; that one understands the world in which one lives.

In order to fulfil this need, people are equipped with very powerful psychological drives. Denial is one of these. The rock group is right. De Nile is not just a river in Egypt. Denial is a process whereby, when someone recognizes a problem or a threat, and with one part of their mind understands it very clearly and with another part of their mind fears it very deeply, a mechanism is activated so that knowledge of that threat is suppressed, kept out of the rational mind. Sometimes when our guard is down, for example when we are asleep, the thing that is being denied can make itself known in our dream life. We will then have a shadowy and worrying recollection of it when we wake up. It may be a sort of safety valve.

Things we deny in this way, by definition we are unable to act against. So denial expresses a great sense of powerlessness. Sometimes the nightmare moves from the night to the day. So it was with many young mothers who were reported by psychologists to be experiencing profound depression and anxiety during the height of the second Cold War in the early 1980s when many people felt that the risk of nuclear incineration was very great. Typically, what these people felt was a sense of complete helplessness in the situation and of total responsibility for their young children. How would a mother in such a circumstance fulfil her responsibility to her child? Gazing at the baby lying asleep in its cradle, people's minds were filled with horror as they wondered how best to kill their children as a last act of mercy to save them from the agony of death from radiation in the event of a holocaust.

Such grim and introverted thoughts are also an example of denial. Like dreams, they serve to deflect energy and attention from the task of actually solving the problem or addressing it head on.

The shadow of the "grey zone"

Nuclear holocaust did not come in the 1980s. But there have been times in the twentieth century when people's exercise of denial in an attempt to protect themselves from dreadful knowledge in the present has left them defenceless. Despite the heroic guerrilla fighting in the sewers of the Warsaw ghetto and several examples of prisoner rebellions in the death camps of the Third Reich, the terrible question which has hung over the extermination of six million Jews in Hitler's genocidal "Final Solution" has been, for survivors and students of that terrifying hell on earth, why, despite the resistance of many, the majority did not resist.

For understanding, we have to depend upon the precious witness of those who went into that hell and yet survived. People like the Italian writer Primo Levi, an industrial chemist, who entered and emerged from Auschwitz and whose searing books give us some inkling of what it was to have been a victim of Hitler's genocidal machine. He records in *The Drowned and the Saved* how people in the process of deportation to the death camps, while they might have some idea of what awaited them, did not discuss it in direct terms. Instead, a whole language of hushed euphemism was used and the assurances of the Germans that people were being taken to resettlement in labour camps, the phoney letters of reassurance that prisoners were forced to write home under duress, and which were sent from some of the camps back to Jewish communities in order to pacify fears, were all too often gratefully believed. Even when they reached the death camps, victims themselves developed and used a vocabulary of euphemisms about what was going on.

Levi describes how the prisoners within the concentration camps (that is to say those who were not killed immediately upon arrival) fell into four distinct categories. There were those who exercised denial by the use of euphemistic language, talking about "Canada" as the destination to which the victims of the gas chambers were going. Then there were the inhabitants of what Levi called the "grey zone". These tragic people also exercised a sort of denial in that they were prepared to collaborate with the camp authorities to act as *kapos* (camp police) or to work in the dreadful *sonderkommandos*, the labour gangs whose job it was to drag the corpses from the gas chamber, to shave off their hair, to extract their gold teeth, to haul the corpses to the ovens, to spread the ash. With their labour, the camps actually ran with minimum need for SS presence. For their collaboration the people in the "grey zone" won some small extension of their own time on earth until either their health or their minds gave out and they themselves fell into the abyss.

Thirdly, there were those in the first category who shortly before their deaths became *mosselmen*, camp jargon for people who lost their minds and became living zombies, thus exercising denial in the most horrendously self-destructive manner by loss of their personalities. And finally there were those who resisted. Some did so heroically but futilely by force, in the several rebellions which did occur in the death camp complex during the period of their operation. Fragmentary news of resistance, like evidence of what was actually going on, was smuggled out to the Allies. But nothing was done. The gas chambers and ovens and the railways leading to the camps were never bombed,

Auschwitz, 1945

although targets close to them were.

Others managed to resist mentally. These were those who managed to retain a strong sense of identity distinct from the depersonalized one which everything in the camp world, from denial of clothes to denial of name (signalled by the tattooed insignia on the forearm) was intended to foster. Generally, such people were those who possessed a powerful religion or a powerful political ideology: Jehovah's Witnesses and Communists for example. These people did not exercise denial in interpreting to themselves the hell into which they had fallen. They knew where they were and so they were able to defend themselves, and, in some cases, to survive to tell the tale.

But for the majority, their fate was literally hellish: hellish in the sense of Dante's Inferno. For they died not once but three times. As Himmler is reported to have said sadistically to a group of victims, "You are about to die and your death will be complete. For not only will you die now, but you have been unable to tell anyone of your fate and if you were able to tell anyone they would not believe you. For once our work is done these camps will be erased. There will be no future sign. You will have fallen into a complete oblivion." It was an invitation to despair.

We ignore at our peril the terrible history of the victims of the holocaust. We owe it to those men and women to learn about the formidable power of denial. Their story teaches us the danger of denial in leaving us defenceless and unable to prepare against a terrible fate approaching in the future but is also – encouragingly – a record of the triumph of the human spirit even under the most extreme horrors. Similar record comes from the Gulag Archipelago. The former prisoner Alexander Solzhenitsyn describes in his novel *One Day in the Life of Ivan Denisovitch* how the prisoner could deny the power of the prison by trying to preserve a private space for himself in his own mind which he refused to allow the prison authorities to erase or to enter.

The danger of denial applied not only to the victims of the Nazi holocaust. It also wrought a terrible effect upon the perpetrators, and this is a lesson which we must heed. In the battle for environmental security, which may not kill with such concentrated intent, but which may kill more people more widely in the end than could the Nazis, we are both victims and perpetrators simultaneously.

The American psychologist Robert J. Lifton has studied closely the Nazi doctors and technologists who played an important role in building and running the killing machine and in developing medical technologies to kill large numbers of people. Within the wider complicit German population, they are therefore the best single group upon which to focus. Their knowledge was indispensable yet, at the same time, they had to protect themselves from the true nature of what they were doing. This they did in three ways by an act of what Lifton calls "psychic numbing". The first way was to "professionalize" their activities. Primo Levi writes, ". . . love for a job well done is a deeply ambiguous virtue. It animated Michelangelo . . . Rudolph Höss, the Auschwitz commander, boasted of the same virtue".

In early 1940, clinical trials were conducted to ascertain the killing efficiency of various forms of carbon monoxide poisoning and the injection of lethal substances (various combinations of morphine, scopolamine, curare and prussic acid). Two leaders of Nazi medicine, Karl Brandt and Leonardo Conti, made a point of personally administering the injections to the guinea pig victims in order to demonstrate the highest "responsibility" to the Führer's order. The gas chamber was operated during the trials by the head of the "Euthanasia" Institution at which the trials took place. Dr Irmfried Eberl also wished to demonstrate his own "responsibility".

The outcome of the trial was to prove the gas method superior, killing people "very quickly ... without scenes or commotion", while the injections caused them to die "only slowly", or in some cases not until a second injection was given. On the basis of this information, Dr Brandt considered the gassing method more "humane" and ordered that "only doctors should carry out the gassings". The ideal of professional medical responsibility had enabled these doctors to pervert their knowledge from the saving to the killing of people.

In more routine roles elsewhere in the Nazi death camps, doctors employed the same psychological defence. Standing on the platforms where the transports of Jews arrived at Auschwitz they conducted "selections", pointing those who required "special treatment" (i.e. gassing) in one direction while others who would be kept alive for some time were pointed in another direction. Professor Lifton observes how the use of the word "selection" with its Darwinian associations, suggests a kind of positive biological activity.

The second way in which psychic numbing was effected was to mystify and elevate the higher purpose towards which these terrible acts were necessary steps. Repeatedly, young doctors being socialized to killing, or to conducting experiments on living prisoners, were told that the brutalization which they had to undergo was heroic and purified their souls, that the "racial death" of the Jews and the advancement of knowledge from their experiments on them were necessary to the triumph of Nazism. It was necessary to kill in order to heal. They could see respected peers and seniors participating and could thus feel reinforced. Lifton finds that the Nazi doctors often used the sense of professional *camaraderie* to overcome any qualms that newly inducted doctors might feel about the tasks that they had to undertake.

Thirdly, the professionals would dissociate themselves from their acts. The people entering the gas chamber were not people, they were vermin, "life unworthy of life". So the feelings and ethical judgements which relate to killing people in civilized societies didn't operate. In addition to narcotics and, in the later stages of the war, mutiny, very similar psychological defences were used by American soldiers during the Vietnam War. Often unwilling executioners, conscripted soldiers depersonalized the Vietnamese as "Slants", "Gooks", "VC (Viet Cong)" or "Charlie", and if not physically wounded, by practising dissociation, they appeared to survive their tours of duty unscathed.

Sometimes, when a buddy was killed, the reality broke through. One deserter, a Marine decorated for bravery who had volunteered for a second tour, interviewed in Sweden explained, "To me it was like on TV. I pulled the trigger and they fell down. But I didn't register that they were actually dead. Not until my buddy had his brains shot out right next to me. I deserted the next day". Of those who finished their tours, alarming numbers of Vietnam veterans fell victim to Post Combat Shock Syndrome in the years after their return. Some might go berserk, like Rambo, but most turned their violence upon themselves, committing suicide or simply dying of illnesses that should not really have been fatal, their constitutions undermined and their energy sapped by the demands of denial.

In his most recent book, *The Genocidal Mentality: Nazi Holocaust and Nuclear Threat*, Professor Lifton carries his analysis of the psychology of Nazi doctors forward into comparison with the state of mind of those associated with research upon and

American soldiers in Vietnam.

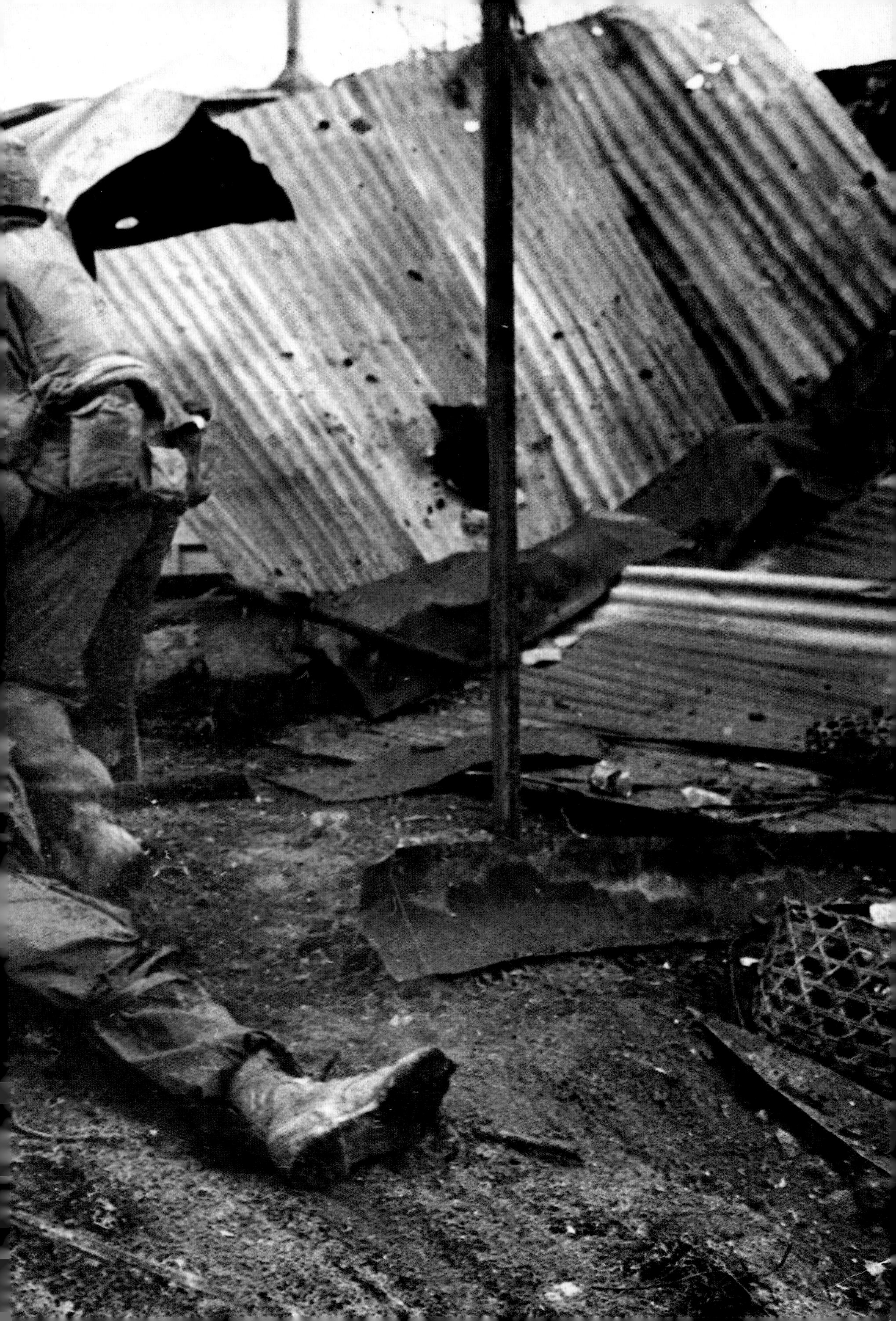

building of the atomic and hydrogen bombs. He finds that many of those who worked at Los Alamos used the same protective devices as did the Nazi doctors. They too felt pride in their work. Robert J. Oppenheimer, who directed the Manhattan Project which built the A-bomb, called the problem, "technologically sweet". They too dissociated their work from the rest of life. The physicist concentrated upon the job in hand. Hans Bethe later wrote, "I had no feelings. I was only concentrating on the job". Equally, many of those who were part of the group that built the bomb referred to the exhilarating, life-enhancing quality of cooperation among their peers upon a task of great technical intricacy. They too mystified the bomb and fitted it within the greater purpose of "deterrence" of what President Reagan was later to call "the Evil Empire".

Denial of what they were doing, in both the Nazi and the nuclear circumstances, had three components: "psychic numbing" characterized by a diminished capacity or inclination to feel and involving separation of thought from feeling; "doubling" – the creation of a second functional self related to the more or less autonomous form of one's other self: the Auschwitz personality as against the homeloving personality, the laboratory personality as against the homeloving personality; and, taken together, numbing and doubling which permitted with relatively little psychological cost sustained actions harming others – "brutalization".

These are harsh and extreme examples, but we use them deliberately because it is here that we can best identify these characteristics which in lesser part we all exercise when, standing in CASSANDRA, we wish to avoid linking cause and effect. We must recognize how deep-seated and powerful are the forces of denial within every one of us, which if we do not confront them, can make us simultaneously both the victims and the perpetrators of acts that can destroy the security of the biosphere.

Professor Lifton ends his analysis on the genocidal mentality by positing that the way to defeat it is by creating what he calls a species mentality. "Species awareness means awareness of human choice: this is not the End of Time – unless we choose to make it so. We need not accept the death sentence . . . we are not powerless." By choosing instead a human future we are, in the words of the Polish Solidarity leader Adam Michnik, "defending hope". In other words we are defending that very thing which Himmler sought at the moment of their death to take away from those about to enter the gas chamber.

What then is the choice before us? What tactic can we adopt? In the terrible circumstances we have used to illustrate this point, the choice between action and inaction, between denial and revolt, has lain entirely in the strength of character and moral fibre of individuals. But it is unreasonable to ask everyone to be a hero. We don't all have what it takes; yet in this battle, we must all take part.

Global Poohsticks

If it is to depend upon the strength of character of individuals alone, then the choice becomes too much dependent upon the higher qualities of the human spirit. It becomes like a game of global Poohsticks. Poohsticks was the invention of Winnie the Pooh in A. A. Milne's classic children's book *The House at Pooh Corner*. Standing on a bridge Pooh and Christopher Robin dropped sticks into the stream, then quickly ran across the bridge to see which stick would appear at the other side first.

In global Poohsticks, one stick represents our response under the constraints of denial and wishful thinking, the other, our response if we can summon up the courage and imagination to face the reality of uncertainty. We drop them into the stream of Time. Which will win?

This is no basis on which to gamble for global security. While it might be laudable to count on the higher qualities of mankind's character alone to carry forward the search for environmental and global security it would not be prudent, especially when we remember the disastrous history of the uses of denial during this century. We have to find ways in which the largest number of people can be helped to deny denial.

Observe! Analyse! Act!

You haven't been feeling too well recently. You have had an ache in your arm for some time but you haven't paid any attention to it and then, suddenly, and rather worryingly, a lump has appeared. What on earth can it be? The thought leaps into your mind. Perhaps it's cancer? If it is, then what do I do? One possibility would be to exercise denial. In fact, there is quite a lot of evidence to suggest that people who suspect themselves to be particularly prone to cancer, those, for example, who live in families with a history of cancer, are particularly loath to be screened when they have no symptoms, or to go to the doctor when they do have symptoms. The logic seems to be that going to a cancer screening centre is tempting fate because you are admitting the possibility that you could have cancer: that admission in itself, and the screening, could in some way actually bring the disease on.

But let's assume that you haven't exercised denial. Full of anxiety, you go to the doctor. The doctor examines you and with a merry smile says, "Oh well, don't worry. I don't think we need bother too much! Here's some cream to rub into the lump and I suggest that you take a couple of aspirins before you go to bed. If it hasn't got better in a few weeks then come back and see me again!"

"But doctor," you say, "what do you think it could be and why are you treating it like this? I'm terrified that it could be cancer. You know my mother died of it."

"Yes, yes," the doctor replies, with an impatient wave of the hand. "Of course it could be cancer, but I think that's jumping the gun a little bit, Cancer treatment is very expensive, you know, and we wouldn't want to waste money on unnecessary investigations and treatments for what may be, as I rather suspect it is, a flea bite!"

"You mean," you say incredulously and with mounting anger, "you mean that you are going to treat this thing on my arm as if it's a flea bite although you admit that there's a possibility that it could be cancer? What happens if you're wrong?"

"Well then," the doctor replies, "in that case you'll be dead and so we won't need to worry about the lump any more, will we?"

If you went to a doctor in such circumstances and had such an interview you would be justifiably angered. If you had any sense you would immediately complain to the General Medical Council about the doctor's unprofessionalism and find another person to treat you. Because when we go to a doctor with symptoms which we and they cannot immediately explain, we expect a certain course of action.

The great success of modern medicine has been partly to do with the drugs and techniques which it has developed, but partly also to do with the analytic process which it demands. The correct course of action is that the doctor does not presume the least significant diagnosis at the beginning of the consultation, but, on the contrary, assumes the *most serious and extreme.*

If the doctor thinks that the lump on your arm could possibly be cancer then the very first thing that he or she will do is to order a battery of tests and examinations which will prove or disprove conclusively whether that worst diagnosis is correct. If it is correct, then time has been gained and treatment can begin which may increase your chance of being able to survive the illness. If, on the other hand, it proves to be a negative diagnosis of cancer, then you and the doctor will regard this as money well spent and he or she will move analytically down the ladder of seriousness to investigate the second possible diagnosis of the lump on your arm. In this way the investigation proceeds until, at the end of the day, he perhaps arrives at the flea bite with which our hypothetical and irresponsible doctor had begun.

The medical model is relevant in another way, too. It, too, "dissociates": it deals with the lump on the arm, not with the total person (at least, not during initial diagnosis). But – the crucial difference from dissociation during denial – once the diagnosis is made, responsible medicine *reintegrates* with the sick person and offers counselling: moral and psychological help in dealing with the illness.

In CASSANDRA, the information on the screens, as Lester Brown observed, offers us a sort of health check on the health of the planet. This is no idle use of words. It is entirely appropriate that the medical model of diagnosis should be used as a principle for action. The view expressed by Working Group Three of the IPCC during the summer of 1990 is not only wrong in fact but wrong in approach. Much to be preferred is the ministerial statement of the Second World Climate Conference published on 7 November 1990:

> We note that while climate has varied in the past and there is still a large degree of scientific uncertainty, the rate of climate change predicted by the Inter Governmental Panel on Climate Change to occur over the next century is unprecedented . . .

Where there are threats of serious or irreversible damage, lack of full scientific certainty should not be used as a reason for postponing cost-effective measures to prevent environmental degradation.

Therefore, standing in CASSANDRA, it is possible for us to use the medical model as a means of producing that integration of understanding of the information appearing on the different screens which the aboriginals lacked and which, in its mind-boggling complexity, might offer us the seductive but dangerous option of denial.

The medical model is reassuring because it enables us to do two essential things. First, it enables us to see the problem in its entirety as a problem of health and, second, it enables us to break up the problem and treat each of the components then identified in a systematic and orderly way. At the same time we would retain an understanding of the potential interaction between the different illnesses being treated in the same way that physicians dealing with a patient suffering simultaneously from a number of complaints keep track of the interaction of the diseases and, of course, the different drugs that are used to treat the different injuries and illnesses. Finally, the medical model does not ignore the need for counselling.

Here the usefulness of the medical model ends. It assumes that once we have made the diagnosis we can immediately find appropriate drugs to hand and, in the phrase of Paul Ehrlich, the father of modern chemotherapy, "fire the magic bullets" which will kill the disease.

But for environmental security, it is much more difficult to move from analysis to action. The danger is that denial could strike again and that we would find ourselves paralysed by our understanding. Lester Brown explains:

It's difficult to see exactly how we will respond to our security in the environmental sense because it will require fundamental changes in behaviour and in values at the individual level and priorities at the national level. Whether or not we can bring about these changes in the time that is available is really the question. . . . We could find ourselves in the situation where environmental the two would begin to feed on each other. Another kind of response is the response of the United States to [the Japanese attack upon] Pearl Harbor [in December 1941] which mobilized the country completely . . . it is not clear at this point exactly how things are going to go but that is what the 1990s are all about.

Lester Brown's choice of historical analogy is interesting. It takes us back to the world of Top Guns as we contemplate the security implications of toxic whales. But Brown is right to do so, for it is in the *military form of analysis* that we can find most immediately, most generally and most powerfully a form of analysis which will help us to move decisively on to the offensive.

You posted your sentries before dusk. The main body of your forces is now encamped and you settle down with your staff officers to try and make sense of your tactical situation. It's always difficult to be a colonel in an isolated position like this. You're a long way from headquarters and the fate of your men and of your unit lies entirely in your hands. You know the enemy is out there and you know quite a bit from your reconnaissance as well as from the stream of intelligence that has been pouring into your field headquarters from divisional HQ. But what is the prudent course of action? What is the safest thing to do?

You review the evidence that you have in front of you. It is possible to work out a rough order of battle. You know that your enemy is short on airpower but has many anti-aircraft missiles. You know that he is short on armour but has quite a lot of artillery. You know that he has many snipers and that your people have been under frequent harassing attack as they have tried to move up and consolidate the fire base where you now sit. Headquarters tell you that the will of the enemy to fight is in doubt. How much credence should you give to that as you sit here on the front line?

The answer is, if you are a competent colonel, very little indeed. The only safe assumptions to make are pessimistic. Prudent military planning at the tactical level involves making "worst-case assumptions".

Worst-case analysis works like this. You have certain information about your opponent's capabilities. You know that he has so many aircraft, so many tanks, so many soldiers. In each case, you assume that the ability of the tanks to work, that the ability of the aircraft to fly, that the numbers of troops involved are better, faster, larger than the evidence in front of you and you plan your strategy to deal with your *assumptions*, not with the evidence on the page.

As to the motivation of the enemy, while Intelligence may be telling you there is doubt about his will to fight, you would be very unwise to assume anything other than maximum cunning, maximum aggression, maximum willpower to destroy you in the opponent facing you, and again to plan your dispositions and tactics to deal with just such an enemy: to assume the worst possible sort of enemy, not to hope for one that is weak and fallible.

Having analysed capabilities and intentions in this extremely pessimistic manner, you must then calculate the percentage risk of a number of different eventualities. Thereafter you must choose how to allocate your limited resources: so-called opportunity/cost choices. With not enough ground personnel and equipment to cover all possible outcomes you must make a list of priorities. First, you cover the most serious and immediate of the risks, proceeding down the list until all your resources have been allocated.

At the higher level of command, the theatre level, the so-called "theatre commander" takes the decision as to where to place complete units and indeed decides the nature of the attack or defence which those units will be required to conduct. In all aspects of war making, safety is maximized by "worst-case analysis".

You can see at once how military "worst case analysis" can be extremely powerful in moving from analysis to action. It forces choices and that, in addressing the environmental security threat, is just the sort of forcing we require. By being so systematic, it can override denial.

But worst-case analysis can be extremely dangerous if misapplied. During the Cold War, the militarization of international relations which came from solving the problem of the Russians with the Bomb meant that the sort of tactical "worst-case" analysis it would be prudent for a colonel to use on a battlefield was also used at the level of foreign ministers. Here, instead of making an attempt to stand in the other man's boots, the presumption was made that the motives of the other side were indescribably bad, that the other side was deservedly demonized, that the capabilities of the other side were awesome, far greater than your Intelligence actually suggested, and, on the basis of such analysis, the conduct of superpower relations was reduced to a debate about whether to build one sort of missile or another sort of missile, one sort of aircraft or another sort of satellite.

Quite rightly, this misuse of military forms of analysis was attacked. It was seen to be a self-fulfilling prophecy because it threatened to make the enemy become in reality the demon that it had been perceived to be. The concept of deterrence, central to western thinking during the nuclear age, was particularly efficient at making this mistake – of carrying military thinking from battlefield to conference table.

Now that the Cold War is over, and we find ourselves moving from a problem of defence against visible enemies to the agenda of environmental security, is the same criticism valid?

Military threat analysis offers the *best and most efficient* way of mobilizing the largest number of people to the tasks of defending environmental security. Because the enemy is not Them in sharp contrast to Us, but instead is within us all and is no other single group, the military ways of thinking can complement the medical model of analysis and help to direct resources at targets identified by the medical model.

The old military way of thinking is efficient at organizing our response to the environmental security agenda in another way. It can help us couple together the different scales of necessary action. Al Gore is a Senator from Tennessee. He sits on the Senate Armed Services Committee and is one of the Democratic Party's most effective spokesmen on environmental issues. He explains:

In looking at the threats to national security coming from the environment, let's use the old military way of thinking. Armies used to talk of conflicts in three different kinds of theatres; local theatres, regional battles and then, during the great confrontation between freedom and communism, strategic conflict and world war. The same is true with environmental threats.

There are local environmental threats, sometimes of most concern to people: water pollution, landfills etc.

Standard procedure for threat assessment

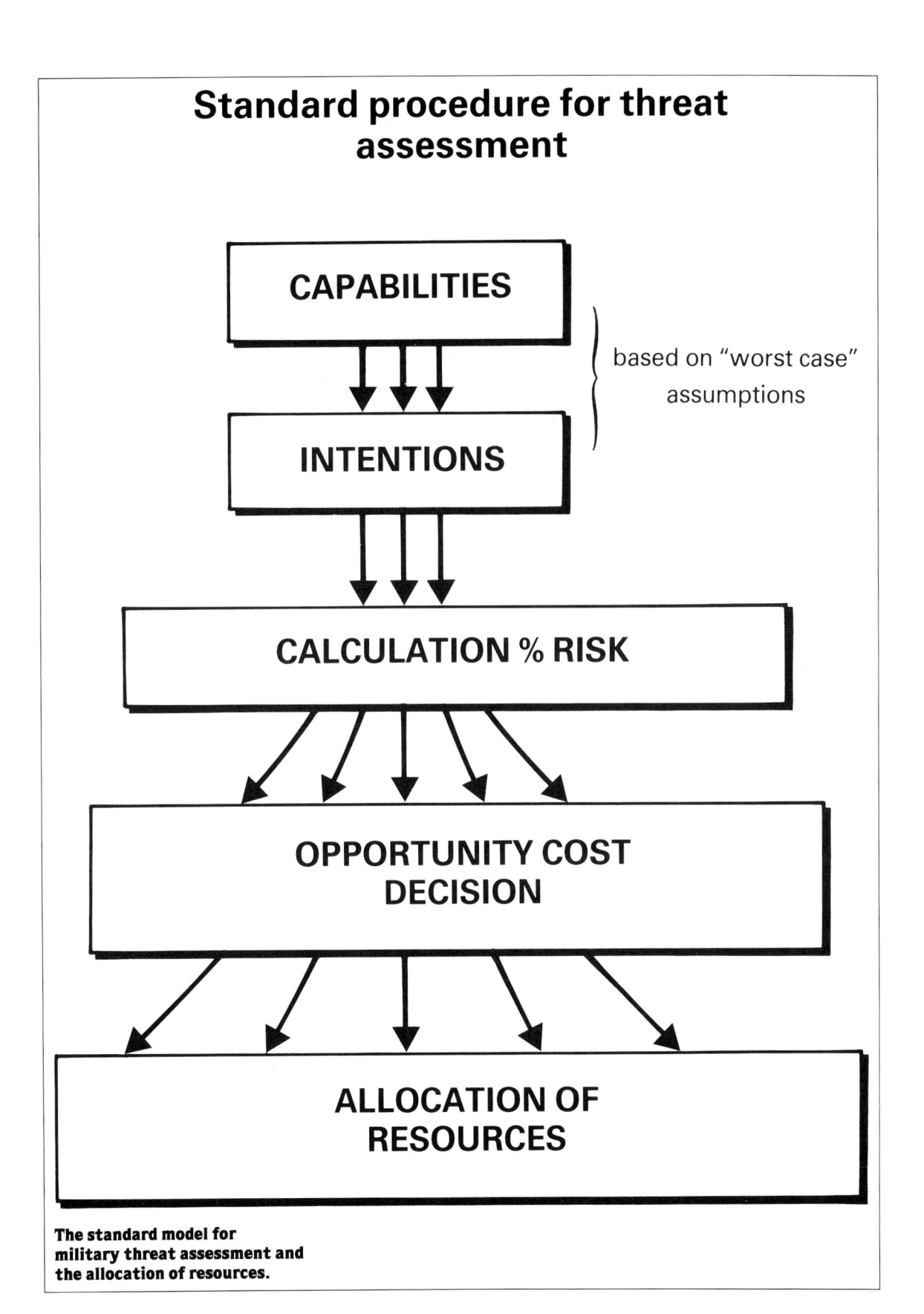

```
        CAPABILITIES
             ↓↓↓            ⎫
                            ⎬ based on "worst case"
         INTENTIONS         ⎭        assumptions
             ↓↓↓

      CALCULATION % RISK
         ↓ ↓ ↓ ↓ ↓

      OPPORTUNITY COST
          DECISION
        ↓ ↓ ↓ ↓ ↓

      ALLOCATION OF
        RESOURCES
```

The standard model for military threat assessment and the allocation of resources.

Then you have regional environmental problems like acid rain, which covers the United States and Canada and ... we have global or strategic environmental threats like global warming or the depletion of the stratospheric ozone layer, or the destruction of the rain forests all over the earth.

Since these new threats are global in nature we have to organize a worldwide response with nations working together cooperatively to meet the new strategic threat.

Nothing enlivens and energizes us as individuals more than the sense that we are part of a collective group or crowd of similarly minded people. As we have already observed, used protectively by the Nazi doctors to strengthen themselves against the true implications of what they were doing, this feeling was also one of the rewards experienced by the young physicists and engineers who worked together upon the atomic bomb project. We need to mobilize it once more in the attack on the new enemies to environmental security, a new kind of security which is the result of shared environmental responsibility.

It is sad but true that we seem to be able to bond in friendship with strangers only against an identifiable threat defined as outside the group of which we are members. Environmental threats pose considerable problems to this way of dealing with the reinforcement of our identities because the source of the threats lies within each one of us. It cannot be externalized. There are no little green men from Mars who are the Enemy. It means that we have to deal with aspects of our own identities and our own behaviour and, as Lester Brown observed, this is a new and perhaps difficult challenge to our way of thinking about ourselves. All the more reason, therefore, to use those parts of our historical and analytic experience with which we are comfortable and familiar and – more to the point – which seem to work well for the type of problem with which environmental security faces us.

Some readers may feel uncomfortable with the sort of proposal that we have just been making, and understandably so. Here we are suggesting that we must prepare to face a radically new security environment; yet we are asking people to carry with them into that future ways of analysing threats that were developed in a past which we hoped to have left behind. Is this not inconsistent? Is there not some ethical problem about doing this?

Senator Al Gore gives the answer:

Some people are concerned that a discussion of environmental crises in language which used to be reserved to military threats to our national security, will bring with it some of the problems that complicated that debate over military threats.

I suppose they have a point. But the advantages far outweigh the disadvantages. For one thing we are used to organizing a total response to threats posed to our national security. We are used to mobilizing the nation's assets and our own thinking and emotions, to confront this kind of threat.

Well, that is precisely the kind of political, personal and emotional response that many of us believe is required to face this environmental threat: a mobilization of assets, a commitment of our personal agendas to help in a common effort to meet these crises. Nothing short of this will suffice. . . . It is that serious.

"A mobilization of assets, a commitment of our personal agendas . . . nothing short of that will suffice . . . It is that serious."

Chapter 4: The Worst of Times

Revolution's choices

It was the best of times, it was the worst of times, it was the age of wisdom, it was the age of foolishness, it was the epoch of belief, it was the epoch of incredulity, it was the season of Light, it was the season of Darkness, it was the spring of hope, it was the winter of despair . . .

Charles Dickens, *A Tale of Two Cities*

In the opening lines of his great novel about the French Revolution, Charles Dickens captured the essence of the revolutionary moment. We live now at such a revolutionary moment. What makes it revolutionary is that perfectly good reasons can be found to suggest why everything should go for the good or that everything should go for the ill. The best and the worst are in violent tension.

For one billion of the world's five billion population, it is already the worst of times. These are the people who live below the poverty line (defined by the World Bank as an annual income of $370 a year). For the world's poor, everything is curtailed: in sub-Saharan Africa, life expectancy is around fifty; in Japan it is eighty. In South Asia the death rate of children under five exceeds 170 per 1,000; in Sweden it is fewer than ten.

A fundamental feature of the lot of the one billion poor is that as their circumstances contract so do their opportunities of reversing them. At a certain point, the poor can no longer help themselves and that duty lies upon the rich. The poor are those who first suffer the consequences of environmental degradation; for the rich use their power to insulate themselves, often at the expense of the poor. Unless the issues of environmental security are addressed, the worst of times for the one billion may well become the fate of many more.

Yet there are many things within us

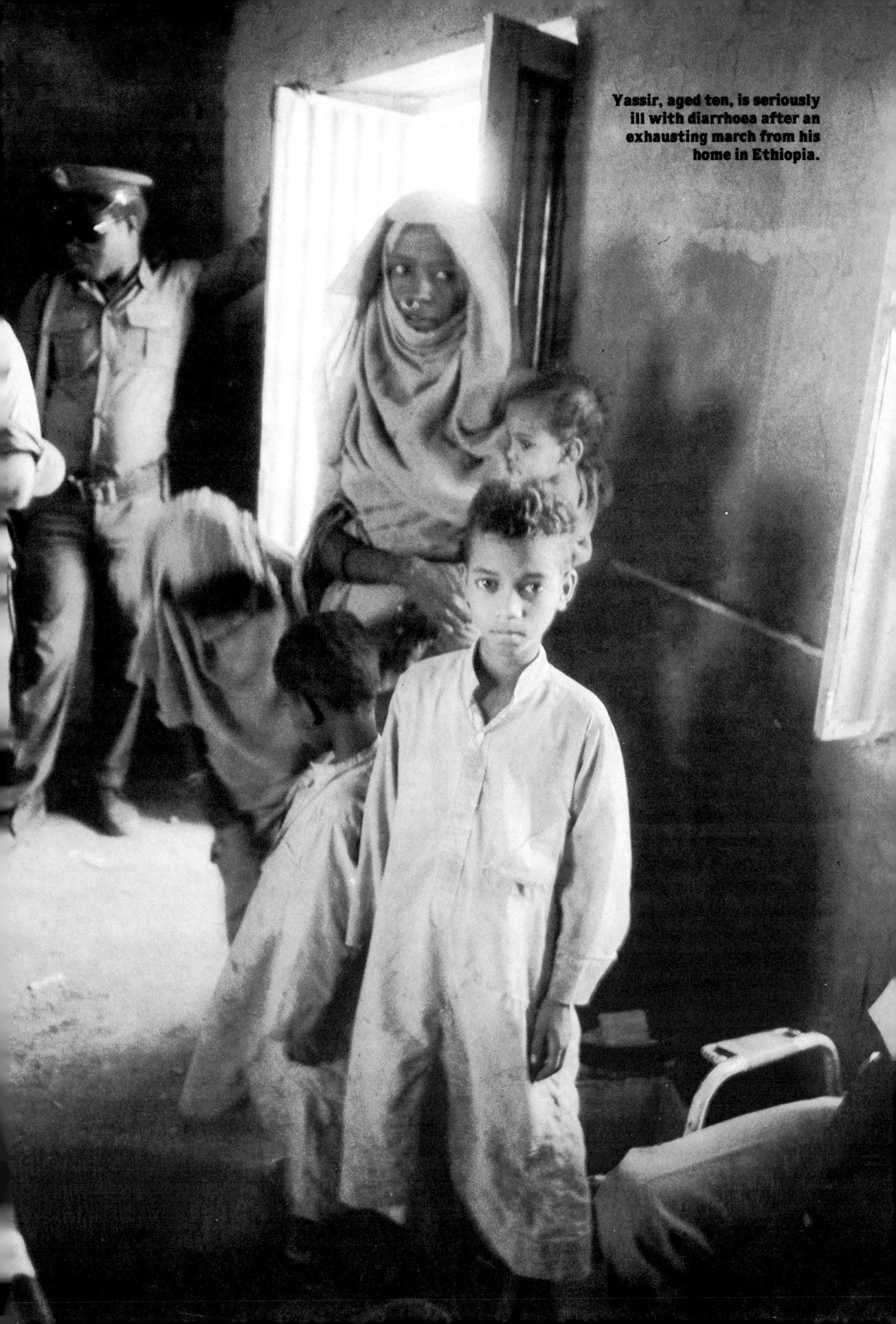

Yassir, aged ten, is seriously ill with diarrhoea after an exhausting march from his home in Ethiopia.

If only one per cent of the world's population were forced to leave their homes as a result of environmental degradation, that would mean sixty million people on the move by the year 2000.

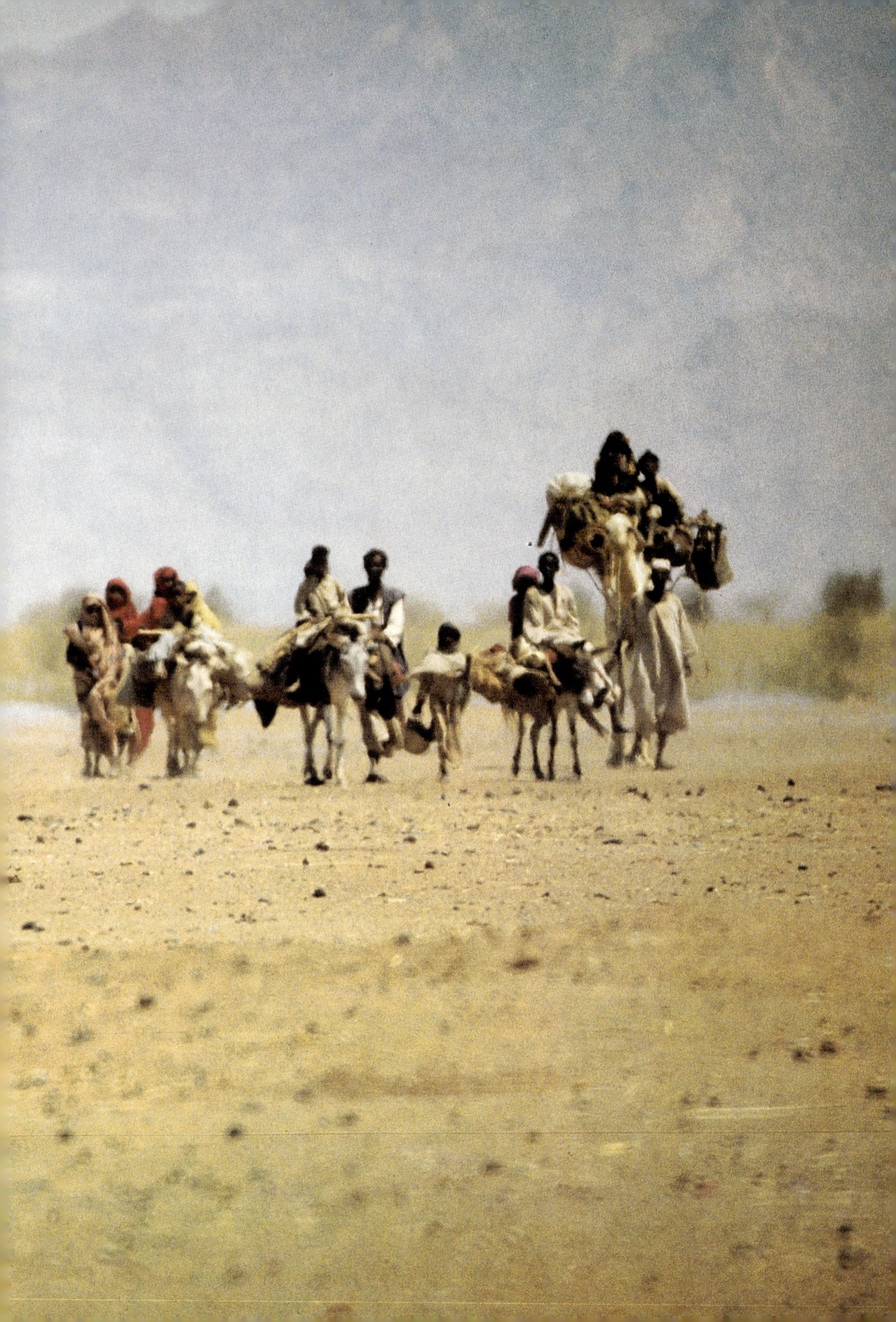

which, in attempting to protect our sense of serenity and of self-identity by denial, serve only to put us further and collectively at risk. If those who do have the choice to make of the revolutionary moment the best of times refuse to exercise it, then the worst of times it may be and it may be so for all. An age of foolishness, an epoch of incredulity, a season of Darkness, the winter of despair. To Dickens's list one should add a failure of imagination and a failure of moral courage.

If it is to be the worst of times what may we expect? In this chapter we shall move from the familiar to the unexpected. We shall move from the private anguish of the individual, left defenceless in a sea of poisons by a fundamental breakdown of the social contract with the state, to the vision of a world filled with refugees as people flee from these and other causes, seeking by moving to find another life. They cannot escape. Many of the environmental stresses, unaddressed and unresolved, will be found wherever they go. Conflict will emerge first in those areas least able to buffer themselves. The final case study in this chapter will imagine what such a war might be like.

Oil Wars

Ask most people to name a natural resource about which they think it most likely that people would go to war and the answer is likely to be oil. Nor would they be wrong. Since the world switched to cheap oil as its primary energy source (a process explained in more detail in Chapter 5), access to oil has become of vital strategic interest. The more a country depends upon oil, the more strategic that interest is. The industrialized countries contain a quarter of the world's population and burn about three quarters of the world's fossil fuels. The United States alone, with 6 per cent of the world's population, consumes a quarter of the world's energy. The USA has burned about four-fifths of its known oil reserves and thus is set to become increasingly dependent upon imports of oil from other parts of the world. Hardly surprising, therefore, that if one is not prepared to contemplate a change in energy-use patterns the only alternative is to find ways in which to guarantee access to that oil at reliable prices and in a reliable and uninterrupted manner.

In 1956, when Colonel Nasser nationalized the Suez Canal, the United Kingdom looked likely to become as vulnerable as the United States will soon be. At that time most of Western Europe's oil was being supplied from the Gulf and from Iran, on the far side of the Suez Canal, and was being transported through the Canal in relatively small tankers. The British Prime Minister of the day, Sir Anthony Eden, had a vision of history repeating itself. As one of those who had argued in 1936 that Hitler's invasion of the Rhineland should be tolerated, Eden reproached himself with failing to stand up to that dictator and the dire consequences which followed. In Nasser, Eden saw Hitler returned and vowed that he would not allow him the same latitude. Partly to protect their oil-supply route the French and British decided to go to war against the Egyptians in order to force them to "internationalize" the Suez Canal. On that occasion the Americans were adamantly opposed to the old imperial powers exercising their muscle in such a military and visible way. Unable to persuade the British and the French by diplomacy to desist, the United States turned to less visible but more powerful means. The Federal Reserve engineered a run on the pound which caused the Chancellor of the Exchequer, Harold Macmillan, to tell the Cabinet that unless the British acceded to the demands for a ceasefire, which was the condition upon which the International Monetary Fund would give emergency special drawing rights to the

British to keep themselves afloat, he would be forced to resign. And so the British and the French had to back down.

In 1967 and again after the Yom Kippur War of 1973, those European countries and America which were seen by the Arab world to have supported Israel (and in 1956 Israel had been an active member of the military coalition against Egypt) were subjected to oil boycotts. Unable to resist the military power of the industrial world, Arab oil-producing countries tried to use the "oil weapon". This was really a political version of the attempts by the Organisation of Petroleum Exporting Countries (OPEC) to regulate the oil price in the international market. OPEC had been founded in 1960 by five countries (Venezuela, Saudi Arabia, Iraq, Iran and Kuwait) after a period of phenomenally successful oil discoveries which had caused the mighty oil companies to decide to reduce the "posted" price of crude oil as a means of stimulating demand, given the vast surplus which they possessed. OPEC was intended to apply a counterbalancing restraint to the unfettered power of the oil producing companies and both in the economic sphere and, after the Arab-Israeli wars, in the political sphere, the ability to use oil as a political weapon proved to be remarkably *unsuccessful*. Unsuccessful, that is, until 1973.

After the 1967 Six Day War countries which were perceived to have been strongly supportive of Israel, like the Netherlands, were targeted for specific boycotts by Arab oil producers. Yet the power of the oil companies which were owned and domiciled in the United States and in Europe (companies like Royal Dutch Shell) was such that by switching cargoes and organizing the distribution system they could successfully frustrate the intentions of the oil producers trying to use the oil weapon.

In 1973 it was somewhat different because the oil majors appeared to have "got in bed with" (in the language of the industry) the oil producers. They now acted more in the interests of the producers than the consumers and, as all the world knows, the oil shocks of 1973 produced a quintupling of price and ended the era of guaranteed cheap oil. The response to this in the United States in particular was to approach the question of oil and oil producers in a somewhat different way. For if they could no longer be controlled through the power of the oil majors they might have to be controlled differently.

The success of the oil cartel in hugely increasing prices after 1973 was resented on two grounds. First, the very principle of a cartel was repugnant to free market enterprise philosophy and second, the rise in oil prices threw the world into recession, perceived as a threat to the security of the United States.

The first response to the oil price shocks of the 1970s was to stimulate, with varying degrees of success, more energy-efficient technologies. The development of energy-efficient technology was, President Carter observed in the 1980s, the "moral equivalent of war". But his campaign quickly faded. One reason for this was that the oil cartel pushed the price of oil *so high* that the rewards of investing in non-OPEC oil energy alternatives were thought to be too risky. After all, if, as was the case in 1974, it cost around $1 a barrel to produce oil in Saudi Arabia and that oil was sold on the world market at a rigged price of over $30 a barrel, to invest in technologies which could not produce energy at less than, say, $20 a barrel was clearly imprudent; for by political or military means it might well be possible to bring the price of oil well below that minimum investment profitability threshold. This is one of the strong components in the case made by proponents of intervention.

The most articulate statement of the case for intervention was made in 1982 by the American strategist Edward Luttwak. "Ever since the 1973 Arab oil embargo, the menace of another and possibly more prolonged and more complete interruption has been a factor in world politics," he wrote. This he found to be unacceptable.

The use of force cannot be desirable; it can only reluctantly be deemed acceptable in circumstances which allow no other choice. It *would* be desirable if oil were afterwards sold at prices which would allow the world economy to recover and resume the progress that the cartel's extortion had interrupted.

At what point, then, would it be *moral*, *legal* and *feasible* to restore certainty in the price of oil to the United States by military means? Luttwak addressed all the issues directly: ". . . the moral issue is defined by a comparison between two sets of consequences: what will humanity suffer if the hypothetical embargo is simply allowed to continue, and what will be the suffering inflicted by an act of intervention designed to restore production and supply?". From this basis, Luttwak weighed the misery that an embargo price of oil of say $100 a barrel would impose upon hundreds of millions of people around the world as against the loss of life in an intervention: ". . . conducted in the empty lands where oil is produced in the greatest amounts, where local military forces are feeble and where no popular resistance is to be expected at all", a phrase which rings rather hollow in the midst of the Iraqi Gulf War.

Arguing from a moral endorsement based upon the greater good of the greater number Luttwak then asserted that an intervention would, in international law, be *legal*. It would be legal because: "If Arab oil producers were to attempt to use the 'oil weapon', weapons of the ordinary sort might justifiably be used to seize, secure and operate the oil fields and loading facilities so long as . . . the embargo would actually deny supplies that could be essential, for which there is no substitute . . .".

Accepting, then, that the intervention proposed would be both moral and legal, Luttwak asked whether it would be feasible. In the light of 1990–91 his answer is interesting. He was in no doubt that, "none of the states in the area have competent armed forces, for reasons so fundamental that they provide a virtual guarantee of immunity for an intervention force." The 1956 Anglo-French invasion was in his view a "text book case of error". A huge armada slowly approaching across the Mediterranean had no element of surprise and therefore the defenders would have had every opportunity to destroy the wellheads and loading facilities and other infrastructure of the oil industry. No, the attack should be sudden and decisive, Luttwak argued. In that case, the response of the invaded oil producing countries might well be to resort to guerrilla warfare or terrorism, he speculated. But all these he regarded as containable problems, especially when set against the ultimate objective of producing reasonable and reliable pricing in the international oil market of the sort which the international oil companies had been able to produce during the halcyon era before the creation of OPEC in 1960.

Following the oil price shocks, the United States established the Central Command (Centcom) and the Rapid Deployment Force (RDF) with the express purpose of training for long range intervention in the oil rich regions of Arabia and the Gulf. Consequently when the élite 82nd Airborne Division landed in Saudi Arabia immediately after the Iraqi invasion of Kuwait in August 1990 it was by no means the first time that that unit had carried out the type of operation for which it had been exercising along with its Marine and Special Force associates for many years. To believe that

Dana Slatkova tests the water in the River Elbe, which has been declared unfit for human consumption.

the massive western military response, and war, led by the United States, to Saddam Hussein's occupation of Kuwait was unconnected to the fact the only substantial remaining reserves of proven, recoverable oil are to be found in the Middle East is naïve, which is not to say that other fundamental issues of international morality are not at issue. It is, however, to ask with Senator Edward Kennedy the question he put in January 1991, in the most momentous debate about war and peace to be held in the United States Senate since the Second World War: whether it was right that "a single drop of American blood should be shed because American automobiles consumed too many drops of oil per mile".

The invisible worm that flies in the night

Refusing to change the physical structure of the American transport system but instead insisting upon a principled "right" of access to the world's natural resources at prices of one's own choosing has heavy consequences. It has meant, in 1991, resorting to violence. In the light of the Gulf Crisis of 1990-91, energy policy based upon "right of access" to other countries' natural resources is very much in people's minds. But for the individual, the worst of times may come by a far more insidious route.

In industrial societies, chemicals are omnipresent. They are used in manufacturing, in the home and have, in the last forty years, totally transformed the practice of agriculture. Indeed, the shift from the "old-fashioned" rotation of animals and plants, which fertilized the soil, kept it in good heart and gave strength to the crops, to rotationless chemical farming, means that really a different name should be used. John Seymour, a celebrated critic of chemical farming, calls it "agribusiness".

Since "agribusiness" breaks the basic rule of husbandry, which is to feed the soil to feed the plant – instead it attempts to feed the plant directly – it is not surprising that by volume, most of the sprays applied to the soil are not taken up by plants but leach down into the groundwater. From there pesticides, nitrates, nitrites, and other agro-chemicals travel into streams, rivers and lakes. In domestic and industrial use, chemicals also find their way into the groundwater, but may in addition be injected as an aerosol into the atmosphere.

From time to time there are horrifying accidents at chemical factories which serve to signal the dangers from this creeping enemy. In 1976 in Seveso in the north of Italy, an explosion at a chemical plant led to a release of dioxin. The Seveso factory was owned by a Swiss company and the accident led to sharp questions as to how it was that the clean, safe Swiss were prepared to put dangerous and dirty industrial processes conveniently on the other side of the mountains. The accident led to the severe disfigurement by chloracne of many children exposed to the dioxin, and the distressing spectacle of dozens of pregnant women petitioning the Pope for permission to terminate their pregnancies: the children that they were carrying might have suffered damage in the womb.

The lessons of Seveso were multiplied many times at the most severe accident yet to occur at a chemical plant. It happened at Bhopal in 1984 in central India, at a pesticide plant producing pesticides for the Union Carbide company of the United States to be sold into the Indian market as pesticides for the "green revolution". A cloud of methyl isocyanate (MIC) killed 2,500 people at once and injured more than one hundred thousand who are still suffering from damaged eyesight, from lung diseases, from intestinal bleeding, neurological and psychological disorders. MIC is related to the First World War gas phosgene, but is fifty times as poisonous.

Spectacular and horrifying as such accidents are, chemicals produced by that scientific philosophy which Rachel Carson described as Neanderthal creep into our lives on a daily basis. In the nineteenth century people lived in a sea of pathogens, surrounded by the threat of infectious illness against which medicine, at that time, had few effective remedies. Today, the triumph of chemotherapy has laid low diseases like tuberculosis and syphilis. Instead we all live in a sea of carcinogenic agents.

There are two frightening things about this. The first is simply that we have to live in this way. Lead, nitrous oxides and the other products of motor exhausts are breathed by every urban dweller. If the weather is right you may have the pleasure of a photochemical smog, while from your food and from ordinary products of everyday living, you may well run risks of cancer which you do not even know about and cannot calculate. That is bad enough. But what is equally frightening is that people learn to tolerate these things.

The great French biologist René Dubos wrote in his book *So Human an Animal* that what distinguished men from rats was that, placed under sufficient stress, rats simply gave up. Lying in a tiny cage bombarded by "white sound" the rat would simply die. But people learn to adapt. This is one of the benefits of our powerful, self-conscious and reflective mind and also one of its greatest dangers. For this is how we exercise denial.

When spring came in the middle of the winter of 1989 to Eastern Europe and revolutions burst out across the middle of that continent, one consequence was the revelation of information about levels of environmental degradation that scandalized, horrified and surprised Westerners. It had been suspected but in its widest embrace had not been fully understood. In retrospect, it should not be surprising that "green issues" should have proved to be so politically potent, so able to mobilize people against regimes which had so flagrantly disregarded the health of their peoples. The former East German regime, for example, had earned considerable amounts of Deutschmarks and hard currency by allowing itself to become a chemical waste dump for Western Europe: a sort of "third world within". Up to unification, West Berlin paid East Berlin Deutschmarks to accept the refuse trucks which brought rubbish from the West into the East.

One area which has suffered particularly badly for many years is northern Bohemia, the former Sudetenland of Czechoslovakia. At the beginning of the century, the Sudetenland was one of the most heavily industrialized areas of Europe, and since then has continued to be, both politically and environmentally, a lost land.

A LOST COUNTRY

Josef Vavrousek is the Federal Minister for the Environment for Czechoslovakia, a distinguished former dissident and leading environmental activist. He was interviewed by Central Television in November 1991.

VAVROUSEK: *Czechoslovakia has to face all the problems of an industrialized country apart from sea pollution because unfortunately we have no sea. Above all we have air and water pollution, which have negative effects on the forests and on the population's health. The problem of hazardous waste is immense too.*
Q. Which are the areas that are worst affected?
V. *Above all the regions where there is the highest concentration of both population and industry. Worst of all north-west Bohemia, an area of 40 x 50 kilometres*

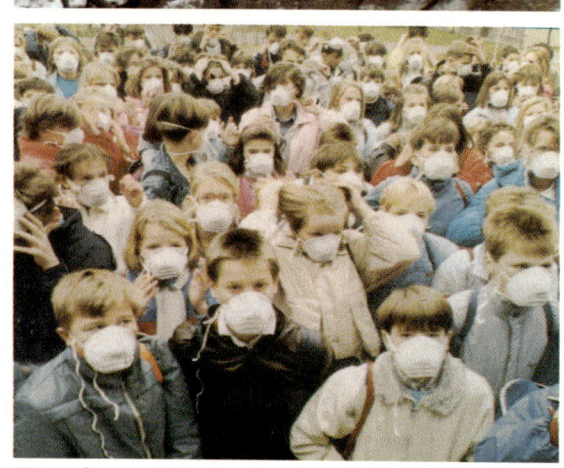

There is an old Arab proverb which says that the most beautiful view of the world is from on top of a horse. Petr Pakosta rides his horse through forests destroyed by acid rain. Forty thousand smog masks have been distributed to schoolchildren in Northern Bohemia, where there is a high incidence of illness related to dangerously high levels of pollution.

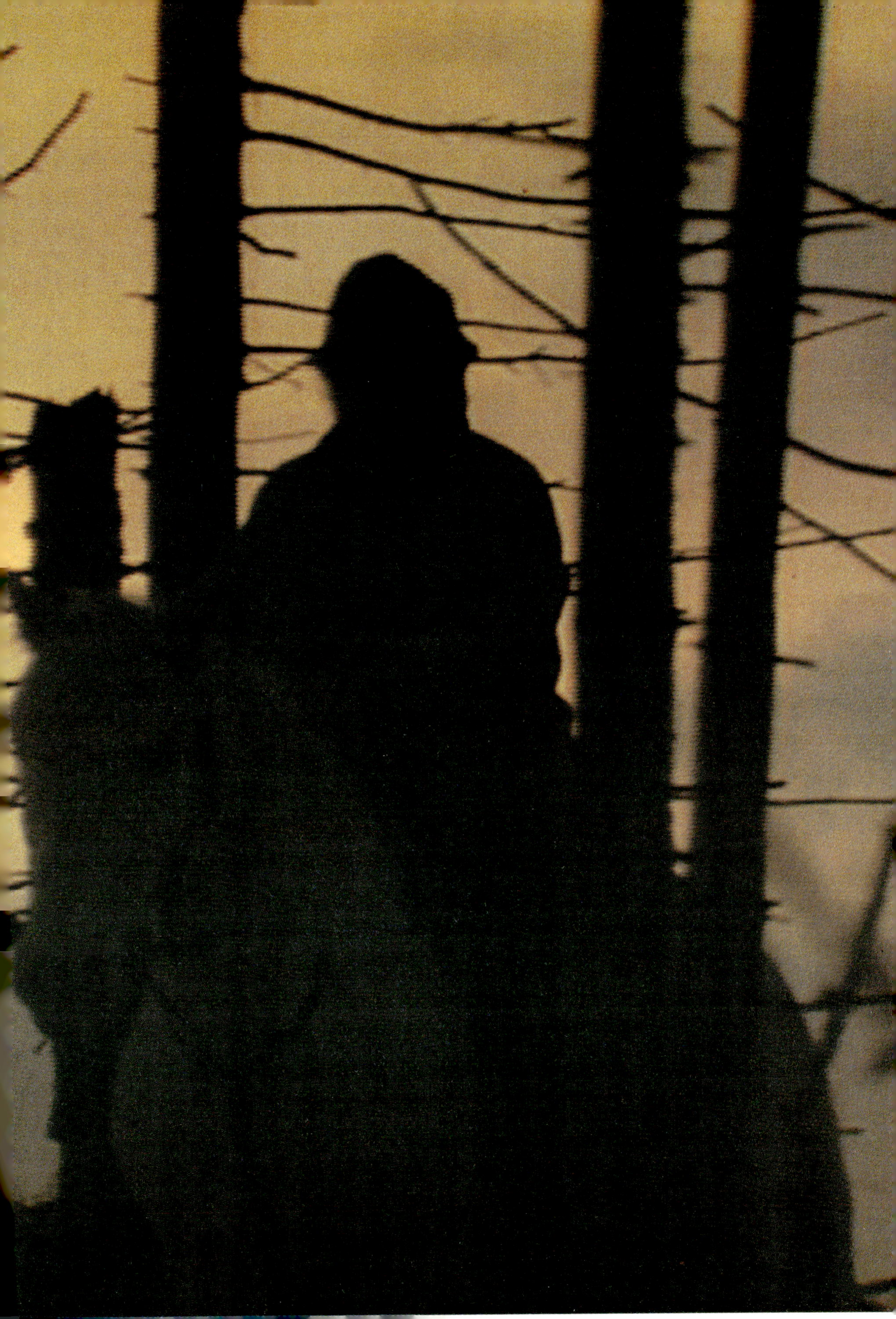

where almost 80 per cent of the republic's energy production is centred and where there is a heavy reliance on lignite coal with its very high sulphur content.

There is air pollution. Frequently for three or four weeks at a stretch the contamination of the air can be as much as twenty times higher than the acceptable norm, which means that for long periods of time people are living in an atmosphere that is totally unacceptable for life.

That leads to a number of psychological problems for people living there who know only too well that they are living in totally unsuitable conditions harmful to their health. They feel that they are being sacrificed because the whole state is creating the wealth (relative as that may be to a western industrialized country) from industry which is ruining their health. It is very dangerous and they are unfortunately totally justified in their concern. The average length of life in this area is as much as five years shorter than the Czech average life expectancy; which in turn is lower than the mean average in West European countries. Unfortunately such psychological problems are reflected in the region's social structure. People with higher qualifications are deserting the region. Therefore the positions which need qualified personnel, doctors, teachers . . . cannot be filled which leads to a further deterioration of the social composition of the region . . ."

Q. What is the most serious problem facing Czechoslovakia today?

V. The most worrying problem is the threat of chaos. Nowadays we can witness it not only in this country but in all of Central and Eastern Europe. After decades of totalitarian regimes – first of all a Fascist one and then a Communist one – came freedom from constraint which in its wake is bringing not only relaxation but also anarchy.

It is a threat to society in general but to the environment in particular. For anarchy and chaos are not conducive to any strategic control. Effort to ameliorate the environment is affected. I believe that now there is too much emphasis on short-term gains, especially in the economy. Too many people are neglecting the long-term consequences taking instead an attitude of carpe diem *(seize the day)*. I feel that the duty of environmentalists is always to be stressing the long term aspects of these matters.

Dana and Stanislav Slatkova live in Usti nad Labem in northern Bohemia, a town surrounded by coalmines, chemical factories and power stations. They are both engineers, working at the local health and safety offices. Dana is responsible for water quality among other things, and frequently obtains water samples from the rivers and tributaries in the area. They have two children, Madelenka and Baborka, and live in a small apartment in a large grey block of flats on a hill overlooking the town.

Dana spoke about her work and about what living in such a heavily polluted environment is like for her and her family as she worked taking samples from the river Elbe.

DANA SLATKOVA: *We are on the banks of the river Elbe (which is no longer fit for drinking water except in case of an emergency). The air is not too good either. Above us, up there above the clouds it is probably sunny, but here in the valley, the fog tends to stay and with it all the smog produced by the surrounding chemical plants. It is not nice here. The inversion usually starts in early autumn, with the worst of all being in January, continuing on into the spring months. So it would be true to say that for nearly half the year we have inversions.*

We have thought about moving away but in today's grave employment situation it is no longer an option. Naturally we worry about the children's health. One daughter is ten and the other seven years old. We always make sure that they spend their long summer holidays away from Usti nad Labem. But in the winter when we could do with a break, we have not got many opportunities to get away from here. So we just sit at home . . . it is simply impossible to go out . . . our children are sick. That is normal here. They are about average. It means flu, sore throats; fortunately our children do not suffer from bronchitis. The older one has eczema, which has been resisting treatment and it is not too pleasant.

Dana's husband frequently escapes to the family's garden up on a hill overlooking a huge open face coal mine.

STANISLAV SLATKOVA: *We bought the plot here to get away from it all . . . ten years ago there was still some fresh air but now look at it. Look at the factories with all their smog and it's getting worse every year. There is very little hope that things will improve in the near future. In the winter quite often the sun disappears sometimes for up to three weeks. That is the worst thing. If one does get out of doors you cannot feel any sun rays. It's really depressing. What is really upsetting is that there is very little one can do about it. Basically just wait and see what will be done. I am afraid that there is very little we can do on our own. I am afraid that it is such a big problem, especially today with all the changes in people's political attitudes.*

Despite the protestations to the contrary, I am afraid that the environment will be relegated in the order of priorities. Economic problems are most pressing. First people need to learn to make money in order to survive and then how to make money which would pay for the improvement of the environment. There is a need to teach people to think in ecological terms. Unless we give the people a clear perspective that ecological matters will be dealt with and the environment improved, I am afraid that within two years people will become apathetic to all environmental issues. What they need is some hope that within a reasonable time, things will change.

Although I don't like saying this, nevertheless I believe that the only solution is foreign aid. The problems are much larger than could be solved within the confines of our republic and the outside world will have to take an interest in this.

Baborka, aged seven, and Madelenka, aged ten, go to school a hundred yards away from their apartment. As they walk down the stairs to their classroom, they pass a reproduction of Picasso's painting *Guernica*, one of the most powerful of all depictions of the effects of modern warfare on a civilian population. It is ironic that it was placed here as a reminder of fascist crimes by the former Communist regime, in an area where the population is under constant threat from a new enemy of that regime's making. Baborka doesn't see it that way. She just reflects on how many of her schoolfriends are regularly ill.

BABORKA: *I would like a nice place with flowers; also animals, fruit and vegetables and some sun.*
Q. Are any of your friends ill?
B: *Well, Lenka Polenkova is often sick, when I would like to play with her, and when I am sick she is fit. Matous Stepan is also often sick and Jitka Vorvelova too. Odra Tisinsky is often away sick. And our teacher, she is also often sick . . . Also Klara . . .*

The list goes on and on. The situation has become so acute that the towns in the valley have distributed 40,000 smog masks with respirators to schoolchildren to wear on the way to and from school on the days when the atmosphere has become so bad that a full-scale smog alert has been declared.

Petr Pakosta who works for a local environmental organization rides his horse in the mountains above the valley, through forests destroyed by acid rain

PAKOSTA: *The devastation started at the beginning of the 1970s. Today we can see the consequences: bare slopes with withered trunks. We have lost roughly 30 per cent of the forests in this area and although we are trying to replace them, the replacement growth has not been successful. In the republic as a whole 80 per cent of trees are threatened and in some regions such as the one we are in now, northern Bohemia, some estimate that 100 per cent of trees are at risk.*

I think that our country can be compared to a person both of whose legs have been broken. It is unlikely that he will mend them himself and learn to walk again on his own. First of all we will need a doctor to help to heal the legs. We will be given a wheelchair. Later on will come the crutches and then some physiotherapy sessions where they will be teaching us to stand on our own feet . . . I am not so much indignant about the environment. What I object to is human indifference and stupidity.

Run for your life!

The Slatkova family felt trapped in their gravely polluted town. Conditions for them were dreadful but not quite bad enough to force them to move regardless. For them, there were economic and other considerations which still mattered. They had not yet reached the point where they felt that they had to run for their lives.

But for many other families around the world, environmental degradation has become that severe; they have had no choice but to abandon their homes and to flee.

There are already places that have become uninhabitable because of environmental degradation. The cause may be industrial pollution as at Love Canal in the United States. Here, in 1978, on the basis of evidence showing a high incidence of reproductive problems among women and high levels of chemical contamination in homes, the soil and the air, the New York State Department of Health ordered the evacuation of pregnant women and children under the age of two from 239 homes immediately surrounding Love Canal, bitterly named.

It may be radiation in the Ukraine in the USSR, where thousands of people will never be able to return to their homes following the explosion of the nuclear reactor in Chernobyl. It may be desertification in the Sahelian zones of Africa. It may be deforestation in the Indian subcontinent resulting in increasing flooding that makes floodplains uninhabitable. In the future, it could be millions of families being forced to abandon their homes as a result of sea-level rise. But that is not yet, although it may be fast approaching.

The present definition of a refugee according to international law is one who:

owing to well founded fear of being persecuted for reasons of race, religion, nationality, membership of a particular social group or political opinion, is outside the country of his nationality and is unable or owing to such fear is unwilling to avail himself of the protection of that country . . .

There were fewer than five million refugees in 1978 and almost fourteen million in 1988. These figures do not include the so-called economic migrants nor do they include those displaced within and outside their own countries by famine, drought, flooding or other environmental hazards (a number estimated at well over ten million people). If only 1 per cent of the world's population of six billion by the year 2000

Fatima is one of thousands of children who have been forced by drought, war and famine to make a temporary home in the Wad Sherife refugee camp in Sudan.

were to be affected by environmental hazards of one kind or another, that would mean sixty million migrants: the population of the United Kingdom. A figure of 5 per cent would produce three hundred million.

In his lecture on environmental refugees to the Natural Environment Research Council in June 1989 Sir Crispin Tickell, former UK Ambassador to the United Nations and Warden of Green College, Oxford, pointed out that when productive land and fresh water are precious no one welcomes intruders and that there is no more ancient cause of conflict between peoples.

"A WELL FOUNDED FEAR"

> Throughout the Third World, land degradation has been the main factor in the migration of subsistence farmers into the slums and shantytowns of major cities, producing desperate populations... (Essam El-Hinawi, *Environmental Refugees*)

Worldwide it is estimated that desertification claims 6 million hectares each year. Africa suffers worst. In the Sahel, an area that stretches from west to east through nine countries from Mauritania to Senegal and on into the Sudan, the problem has been at its most acute. In the droughts which affected the Sahel between 1982 and 1984 one hundred and fifty million people in 24 countries were on the brink of starvation and by March 1985 more than ten million people had been forced to abandon their homes in search of food.

North of the Sahel, as well, growing populations of people and herds have exceeded the carrying capacity of arid lands in Algeria, Egypt, Libya, Morocco and Tunisia. These are the very areas that Professor Martin Parry and the expert group of the IPCC which looked at the agricultural impact of climate change identified as being most at risk: semi-arid tropics. They are strange and difficult lands. The fragility of the balance between plants, soil and animals is great and the skill of nomads and agriculturalists who found ways of living in balance with this harsh and unforgiving environment was greatly to be admired. Yet their ability to maintain that balance has been broken down. Frequently that breakdown has come as a result of choices, which, under stress of population increase and a failure of imagination we may all face in the worst of times. It is called the Goatherd's Dilemma and it goes like this:

You own a herd of goats. You live on the edge of the desert and your goats browse widely on the thorny scrub. Every evening and every morning they congregate at the watering hole. Last year in your herd of one hundred goats you lost fifty from drought and from starvation. Many more goatherds now try to use your watering hole. The tracks of the animals and the weight of their demand upon the thorny bushes surrounding the watering hole have produced an ever wider circle of complete destruction. It's like a ripple effect moving out from the now dried pool into the surrounding countryside. What are you to do?

You need goat's milk for your children, and meat and skins. Last year you lost half your goats. So the sensible thing to do is to try as quickly as possible to double your herd size. That way, if you lose another half, you will at least have more than fifty left.

This is a perfectly rational line of argument for the goatherd in the short term. But, of course, making a decision that seems rational in the short term acts fatally against his interests in the medium and long term, for as each goatherd responds to increasing environmental stress by increasing the size of his herd, so the stress upon the environment accelerates and the collapse of the whole ecosystem advances. His "sensible" short-term strategy turns him into a refugee.

FATIMA'S MARCH

Sitting by a fire in Wad Sherife, a vast refugee camp in western Sudan just across the border from Ethiopia, Fatima, aged ten, tells the story of her family's flight from their home.

FATIMA: *I woke up in the morning and saw my parents loading the donkey. The hardest part of all was when we had to walk because the donkey was so tired and then it had to rest. We tried to stay under trees while the sun was really hot and walked during the night. On the first night we went to a forest. We spread out our clothes and went to sleep. We woke up at dawn. We only had dates to eat and we had to be careful with those. We walked for six days.*

Her father Ibrahim explains why the family had to leave:

IBRAHIM: *This year there was very little rain. The crops sprouted but then they died. I had a goat, which I milked for the children. Then I had to sell it. When the money ran out we left.*

Mariam, her mother, told how she was simply glad that they had been able to take the children away from drought and war and that now they were in the camp they would stay there and "pray to God".

Wad Sherife squats in a barren plain dominated by mountains in the distance. This dismal place first hit the headlines during the droughts in 1984–6 and at the height of that famine it was home to over one hundred thousand people. In 1990 when Fatima and her brothers and sisters arrived, there were fifty thousand people there, many living in makeshift tents made of a tarpaulin and sticks. Up to three hundred people were arriving each day, exhausted from their journey and many suffering from malaria, measles, pneumonia and tuberculosis.

As the camp superintendent at Wad Sherife points out, the refugees who come to Sudan have left poor countries only to place an almost intolerable burden on another poor country. The pattern is repeated all over Africa where poor farmers, both pastoralists and agriculturalists move from one degraded area to another; the destitute placing an even greater burden on the poor.

The search for a better life also propels refugees towards the rich world, and this too can cause tension and conflicts. In the BBC's film *The March*, first shown in 1990, which was set in the near future in the Sudan, after climate stress has further eroded that country's agriculture, a charismatic leader, The Mahdi, tells the EEC Commissioner for Refugees, "In your countries, you spend more on feeding your pets than you do on feeding us. Let us come to you and be your pets. We shall sit by the fire and purr. Or if you refuse, then we shall die in your streets, in your sight, not invisible here!" So begins a great march of the destitute across North Africa to Gibraltar where the film ends with The Mahdi and his people confronting the blue-helmeted troops of the United States of Europe, placed there to keep them out.

But this isn't just a fictional fear. On the Mexican/American border, the makings of such a confrontation are played out daily.

TO REACH EL NORTE!

The degraded farmland near Oaxaca in southern Mexico is a place from which thousands of farmers have been forced to move in recent years. A schoolteacher, Fransisco Ceera, returns home. He now lives in Tijuana on the Mexican/American border where he runs a school for the children of his fellow Mixteca Indians who have moved away from land that they cultivated for centuries. However, many of those who have made their way north to Tijuana are here as a staging post on a journey which they hope will take them on into the United States. The story begins in the village of Tonala about five hours drive west of Oaxaca.

THE MAYOR OF TONALA: *What has been prejudicial to us is that it hasn't rained like it used to. The erosion in San Sebastian has washed away all the [soil]... A lot of people have left the town. They go the state of Zinaola, Baja California and many go to the United States because the harvests are not now enough to live on... what is cultivated is not even half the hectarage that was cultivated... ten years ago. There are no [other] sources of work. There are no factories... The land is the only source...*

FRANCISCO CEERA: *I went to Tijuana because someone was needed to look after the migrants' children from my area which is the Mixteca area around Oaxaca... and now I am headmaster of the school... There is a lot of vandalism and vice here so one has to be very careful... At the border things are very difficult for the young and a lot of them get lost... there is a lot of vice in the gangs... I miss my birthplace, my home town, my family.*

This is the Mixteca land, known as the land of the sun. Many years ago the ancient Mixtecas began to work the land... There was plenty to go round with a lot of vegetation but with the passing of time, it began to erode... They cut down trees without a care... The small cattle rangers rented the land so that they could graze their animals and, in the long run, the land was very impoverished. Now the pasture is finished... and the rain cycle has become irregular...

I now see my homeland with sadness... There is no reafforestation to protect the little amount of earth that is left... In two or three years from now it will be finished... people work only with shovels and spades. They have asked for help but it hasn't come... We spoke to the governor only two days ago and he said he was going to send the machines so that these people would have support. I hope that we are going to get it... they are good people.

So times have been constantly changing but the people's beliefs carry on... in the way of their forefathers. They have always adored all their fields and the land... On the first of April and May there was a ceremony. There was a belief that one should talk to the water, to the rain. In other places people would go their lands carrying a "pulque" (a strong liquor) as a present, talking to the four winds, to the air so that it should be productive and that there should be enough rain... There were deep held beliefs but with modernization and the severe climate change, the lack of rain and social changes that have occurred, these things have disorientated the people and they have stopped working their lands and little by little they are abandoning their lands and are going to live in the cities... there is not any confidence left in the land... in this region what are people going to eat? So who is now going to sow the crops? If they are going to eat vegetables who is going to look after the vegetables? Where are they going to come from if the soil has been washed away?

In Tijuana a Mixteca farmer is waiting to cross the border into the United States.

FARMER: *I come from San Juan Onotapec.*
Q: What are you going to do to get to the USA?
F: *I am looking for a coyote [guide] to take me.*
Q: Do you know how much the coyote will cost?
F: *No. I don't know exactly. But it will be three hundred to six hundred dollars, more*

Decades of misuse have led to severe soil erosion in the Oaxaca region of Southern Mexico.

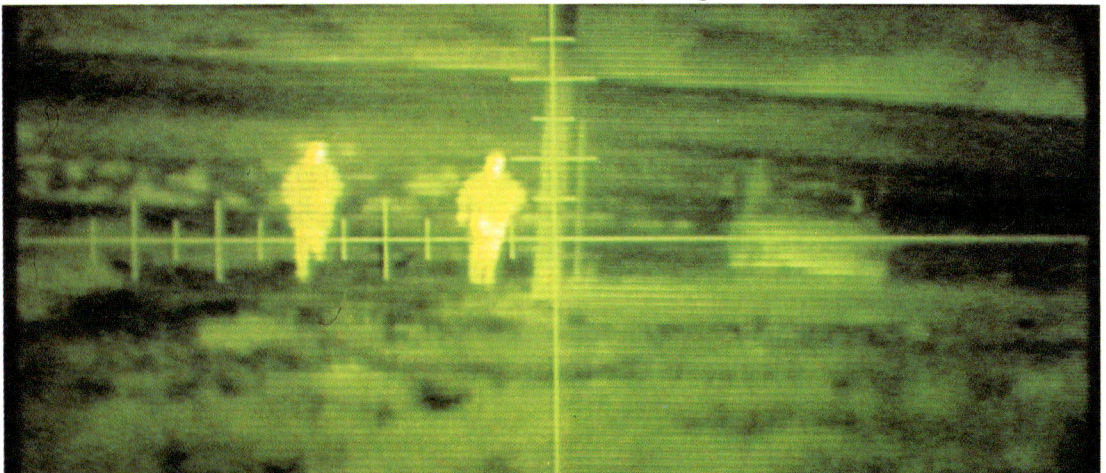

Fransisco Ceera (*left*) is a Mixteca Indian who now lives in the border town of Tijuana, where he teaches the families of fellow Indians who have been driven off their land by a changing climate and soil erosion. Although manpower is stretched very thin, the Border Patrol (*above*) has a considerable array of technology at its disposal.

or less.
Q: Have you got everything planned out?
F: *Yes I have got it planned. I just have to wait until January in order to get across... I have a cousin living in Chicago. He is going to send me some money here.*
Q: Are you going to Chicago?
F: *No, to get there it will be another six hundred dollars*
Q: Do you know where you are going to cross?
F: *Well, at the moment I don't exactly where... the coyote will know...*

Another man waiting to cross reflects the determination of many to make their way into the United States.
MAN: *I crossed and took the train to San Diego and it was there that the immigration police got me.*
Q: They deported you back to here? Are you going to try again?
M: *I am going to try again. We will see who gets tired first.*

On the American – Mexican border a nightly game is played out between the would-be immigrants and the American border patrol. Over the last four years, the border guards say that they have been apprehending many more women and children. Originally the "illegals" as the border patrol call them, were single men seeking to earn money to send back to families back home. Now the whole family comes. They are all vulnerable to bandits and the desert is cold.

The United States patrols a long and difficult border with Mexico, which passes through wild country. To do it, helicopters and military sensors are used as well as border guards in jeeps. Here Kim Jenkins, one of the border patrol officers, speaks on 20 November 1990:

JENKINS: *The unit that I am supervising consists on any given night of six agents and we have been averaging about a hundred and fifty apprehensions in an eight hour period... About a year ago me and another agent with the help of the helicopter rounded up a group of a hundred and seventeen [men, women and children] who were all from the same town in Mexico. They come from any place that you can think of in Mexico and from other Central American countries too.*

Standing on the levee, a brightly lit "no man's land" between the American and Mexican borders, he points out some "illegals" waiting their chance: *There's some people up there waiting their opportunity to make their way across even though it's brightly lit. They'll still try it. They are waiting to dash across the border. It's only about a hundred and fifty yards or so... On the other side of the levee in San Ysidro [on the American side]. They hit the city streets and blend in with the population and it makes them difficult for us to find them. They are very patient. They will wait all night until they get their opportunity...*

Living here in San Diego County and listening to the news and knowing what's going on here in this particular part of the country, they have become a drain on some of the social welfare systems and there's a lot of county administrators and state administrators in the State of California that resent that; and being a taxpayer in the state of California I kinda resent that too. But on the other hand, I'm employed, and it's my job to apprehend these people when they come into the United States. Personally, I feel that they are just looking for a better life and they're doing what anybody else would have to do to make ends meet... If I was in their shoes I would probably do the same thing.

In industrial countries many people feel that limits have to be set on the number of refugees allowed in, however deserving they may be. The dilemma faced by the

British government over the Vietnamese boat people in Hong Kong shows just how difficult these issues are. The rise of overtly racist political parties in a number of Western European countries bears witness to the nastier end of such concerns.

A report by the European Parliament drawn up on behalf of the Committee of Inquiry into Racism and Xenophobia in August 1990 offers an unpleasant narrative of steadily increasing numbers of racial clashes throughout the 12 member states. The report was concerned about conflicts sparked by racial and religious differences: between white Europeans and non-white immigrants and Muslims. It also looked at tensions reflecting competition from skilled East Europeans. In France, where the immigrant population is 4.5 million (7 per cent of the total population), the number of racist attacks has increased markedly and tension between North African Arabs and the police is running high. Anti-Semitism raises its ugly head again all over Europe. The report claims that as many as seventy thousand racial attacks occur each year in Britain.

In southern California, the "Light Up The Border" campaign in San Diego is at the less violent end of the spectrum of white protest. It is only designed to intimidate. Dozens of vehicles drive to an area favoured by the immigrants for attempted crossings and, as the daylight fades, they turn on their headlights. At the other end of the spectrum lurk the pointed hoods and blazing crosses of the white supremacists. The Ku Klux Klan still rides out.

Vietnamese boatpeople and Mexican wetbacks are a small-scale warning of what we might expect in the worst of times.

Water wars

Of the world's water 97 per cent is salt water. Only 2.59 per cent is on the land and most of this water is not available to us because it is locked up in snow and ice at the Poles. Only a tiny amount, 0.014 per cent of the earth's water, is freshwater in free form, available for plant and animal life.

Put another way, the hydrological cycle of rain, evaporation and vapour transport carries about 385,000 cubic kilometres per year. After subtracting the rain and snow that fall directly on to the oceans, flood run-off, run-off in uninhabited areas, absorption by vegetation and the counterbalancing atmospheric vapour transport from sea to land, about 9,000 cubic kilometres of freshwater is available for human use. This is enough to sustain 20 billion people.

However, freshwater is unevenly distributed. Most countries in Africa and the Middle East, much of the western United States, parts of Chile and Argentina and most of Australia suffer water shortages. Between these countries, there is substantial variation in per capita consumption. The average US resident consumes more than seventy times as much water every year as does the average Ghanaian.

The heaviest demand on freshwater supplies comes from agriculture, which accounts for three-quarters of human water use. Three million square kilometres of land have been irrigated, which is an area nearly the size of India, and more is being added each year. As demand grew in the 1960s, the fashionable first response was to deploy "macho" technology by building dams or by tapping groundwater to get water. In contrast, countries like Israel have shown in the Negev Desert that much can be achieved by applying careful thought with each drop of water, increasing the efficiency of water utilization. Increasingly, many Middle Eastern countries draw water from the sea through desalination plants. Another alternative to a dwindling or limited supply of water for domestic agriculture is to import more food.

But irrigation is no permanent route to

increase food production, for salinization is a serious threat to irrigated soils, making them "time-limited". When water collected from rivers by dams evaporates or is absorbed by plants, salt residues build up in the soil. If the rate of deposition exceeds the rate of water-flow able to wash the salts away, the residue accumulates. Irrigated lands in arid regions are especially at risk. More than a million hectares a year worldwide. In the United States alone 20 per cent of irrigated land is affected by salinization.

So the water crisis is both of supply and in the nature of the use of water. There is an increasing problem in water supplies in 80 countries, directly affecting two billion people. In India out of 2,700 water wells provided by the government, 2,300 have dried up. The water table beneath Beijing in China is sinking by up to two metres a year. The Ogallala Aquifer in the USA is the largest known in the world and at present it is being depleted at a rate ten thousand times faster than nature replenishes it. Texas has already used a quarter of the aquifer's water for irrigation. In the Soviet Union, the heavy demands of a bloated cotton industry have led to the ecological destruction of the Aral Sea. Lake Baikal, the largest freshwater lake on the planet, is threatened with pollution. The global use of water doubled between 1940 and 1980 and is expected to double again by the year 2000.

Shortage is endemic in North Africa and competition for the waters of the Nile, the Jordan, the Yarmuk and the Euphrates is intense. The potential for "upstream/downstream" conflict between states is great, especially when water conflicts come on top of other causes of hostility. The Euphrates is a particularly serious worry. Turkey, the upstream state, is currently in the process of building the Ataturk dams which, when complete, could so reduce that river's flow that Syrian and even more so Iraqi agriculture could be severely disrupted. We may confidently predict that neither country will tolerate living with such a threat hanging over their heads in the long term; whereas Turkey has plans to become the regional superpower. Even with prior announcement, when the flow was reduced in 1990 to fill one of the lakes, the effect was dire. At present Turkey and Syria oppose Iraq in the 1990–91 Gulf War alliance.

Of the major river basins, 214 are shared by more than two countries. Twelve river basins are shared by five or more countries.

> The Danube flows through 12 states
> The Niger flows through 10 states
> The Nile and the Congo flow through 9 states
> The Zambezi and the Rhiné flow through 8 states
> The Amazon flows through 7 states
> The Chad, Volta and Mekong flow through 6 states
> The La Plata and the Elbe flow through 5 states.

Disputes over water use and quality simmer in virtually all parts of the world. An estimated 40 per cent of the world's population depends for drinking water, irrigation, or hydropower on those 214 major river systems.

Freshwater may prove to be a more combustible commodity than oil in the future. Certainly, the CIA is prepared to consider that possibility. This map is produced by the Agency and speaks for itself. It shows disputes between riparian neighbours over access to rivers as one of the

> **Lieutenant General Henry Hatch (Officer Commanding United States Army Corps of Engineers):** "There is a strong possibility that environmental degradation could provoke armed conflict in the military sense . . . if it exacerbates the economic differences between the haves and the have-nots, that has been a traditional source of conflict . . ."

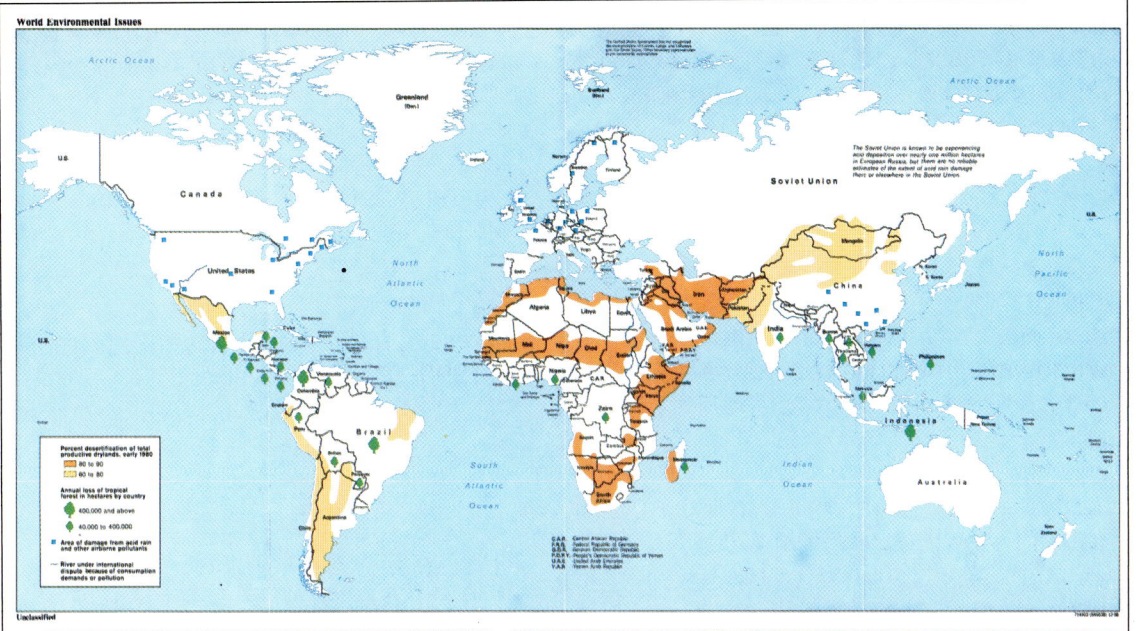

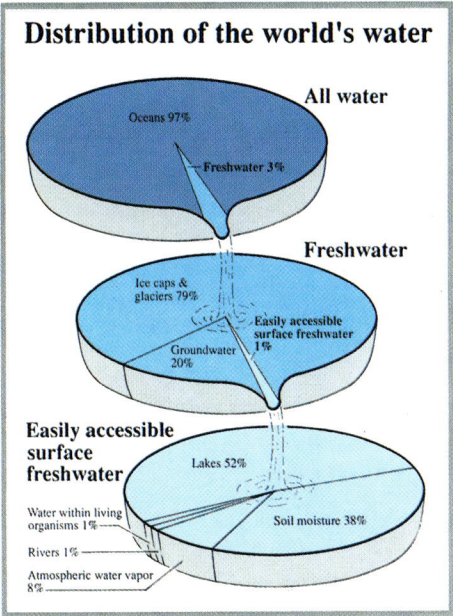

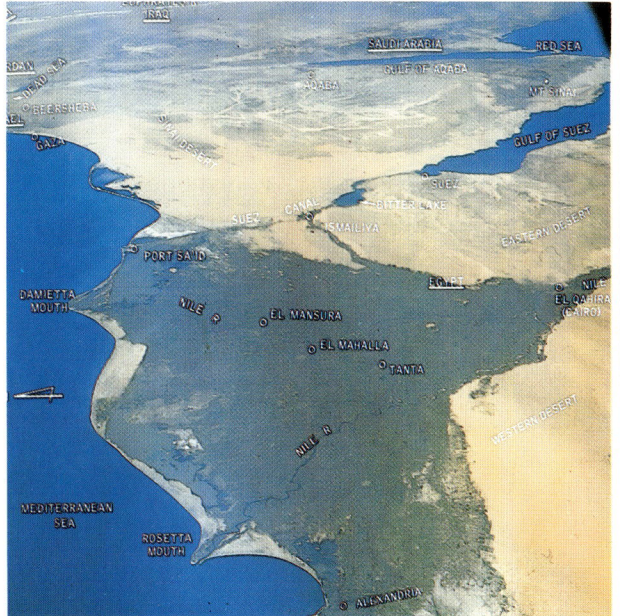

The map (*top*) highlights the areas and issues considered most likely by the Agency to be sources of conflict over natural resources. This photograph (*above right*), taken from a hot air balloon, clearly shows the boundary between the two per cent of cultivated land where ninety eight per cent of the population lives and the ninety eight per cent of the country which is desert.

significant environmental threats to stability and security in the future for which the old fashioned military security agenda believes that it must now prepare.

OPERATION PHARAOH

Background briefing

"For my ancestors the Nile was a god, and I believe that for the Egyptian, the Nile is life," says Butros Ghali, the Egyptian Foreign Minister, in October 1990. Hardly surprising, when most Egyptians are concentrated in the green ribbon of the Nile Valley and the Delta: 2 per cent of the country. Of Egyptian territory 98 per cent is a barren desert. The Nile is the longest river in the world. The White Nile flows 5,600 kilometres from its source in Lake Victoria, the Blue Nile, 4,500 kilometres from its source in Lake Tana in the Ethiopian Highlands to the Mediterranean.

There is no agreement on uses of the Nile waters that binds all the riparian states, but in 1959 an agreement was made between the Sudan and Egypt which fixed Egypt's needs at 48 billion cubic metres of water annually and Sudan's at 6 billion. An increase was agreed following the construction of the Aswan High Dam, which provided an 7.5 billion cubic metres to Egypt and 14.5 billion to Sudan. One hundred billion cubic metres of water leave Ethiopia each year, carrying with it also silts which have, over millennia, helped to form the rich soils downstream.

In 1978, President Anwar Sadat summed up the Egyptian situation very simply. "We depend upon the Nile 100 per cent for our life. So if anyone, at any moment thinks to deprive us of our life we shall never hesitate [to go to war], because it is a matter of life or death."

In 1980, he restated the issue more forcefully. "We do not need permission from Ethiopia or the USSR to divert our Nile water. . . . Tampering with the rights of a nation to water is tampering with its life and a decision to go to war on this score is indisputable in the international community."

In 1985, Butros Ghali echoed Sadat's opinion. "The next war in our region will be over the waters of the Nile, not over politics." Reflecting on that statement in 1990 in the midst of the Gulf crisis, he did not retract:

> Maybe it was a pessimist interpretation of the future of international relations in this region. But I was talking even before the continuing droughts and the problem of desertification. The facts are that in the next few years the demographic explosion in Egypt, in Kenya, in Uganda, will lead to all those countries using more water; and unless we can agree on the management of water resources we may have international or inter-African disputes; and that is why I was saying that the real problem in the next twenty years will not be a problem of boundaries or even of ethnic disputes but a problem of water.

Considering the potential for conflict over water resources, the example chosen by Senator Al Gore was the Nile:

> If you look at the nations bordering the Nile, all of them have rapidly increasing populations. Egypt has 52 million people today. The best case analysis is that within thirty years they will have one hundred million people . . . well the Nile does not have any more water in it today than it did when Moses was found in the bulrushes. . . . If you double or triple or quadruple the number of people seeking access to this limited supply of water, at what point does that create conflict?

Scenario
(the year is 1998)

1. The Nile River Authority tells the Egyptian government that sedimentation because of reduced flow in the Nile is even

worse than they had thought. They have also noted a small rise in sea level and increased salt ingress into the Nile Delta, spoiling fertile land. This, some experts believe, might be an early indication of global warming.

2. In February 1998 a unilateral decision is taken by the Ethiopians to construct a dam on the Blue Nile.

> Background: The Ethiopian civil war has ended with the overthrow of the Dergue and success for the Eritrean and other regional national movements. A new, federal western-inclined Ethiopian state is created. The Federal Coalition now seeks a rapid reconstruction of the country. They accelerate the policy of moving people from the degraded highlands of Welaga province to the more fertile areas of the Blue Nile headwaters. Ambitious development plans include the Blue Nile dam – involving the Japanese as financiers and engineers. This would divert substantial amounts of water from the main Nile Valley. Their development plans receive broad international support from the FAO (the UN Food and Agriculture Organization), the World Bank and the world community after the years of terrible famine in 1972-4, 1984, 1990-91 and 1993-4.

3. The civil war in Sudan, which began in 1983, has led to victory for the southern resistance movement and the government based in Khartoum has been overthrown. Development in southern Sudan is taken seriously, and the new government decides to exploit the southern headwaters of the Nile.

4. Because of the scale of their proposed development plans, the new Sudanese government seeks to reduce the water allocation to Egypt from the Jonglei Canal. Peter Gleick, an American water security expert, explains the background:

> Many consider the Sudan to be perhaps the greatest potential bread basket in Africa. There is a tremendous amount of arable land there that if irrigated could grow crops. Were a serious attempt to be made to do that, Sudan could greatly increase their water consumption and take a lot more water out of the Nile.... The White Nile loses much of its water to evaporation in a region called the Sudd Swamps and there have been plans for many years to build a bypass those swamps and to take water out of the White Nile north. This is the Jonglei Canal. Construction started on it many years ago. An agreement was reached between Egypt and the Sudan to share any additional waters that the Jonglei Canal would save. Construction stopped during the civil war and the unrest in Southern Sudan. If the Sudanese were determined that they needed more of the added flow that the Jonglei Canal would provide we could see a situation where the additional flow is not delivered to Egypt."

Diplomacy

5. The Egyptians demand a Nile Summit to seek a general regional agreement for the management of the Nile in which all riparian states participate. It is held in Geneva in the summer of 1999. Egypt proposes a comprehensive schedule of water abstraction quotas. Sudan is anxious to fulfil its potential and become a major grain producer for the region. The Ethiopians remind the conference of the famines of the last twenty years and argue that the Blue Nile dam is their best defence against recurrence. The Ethiopians point out that they have never been party to a Nile agreement and see little point in starting now.

The meeting becomes tense. The Ethiopians walk out, accusing the Egyp-

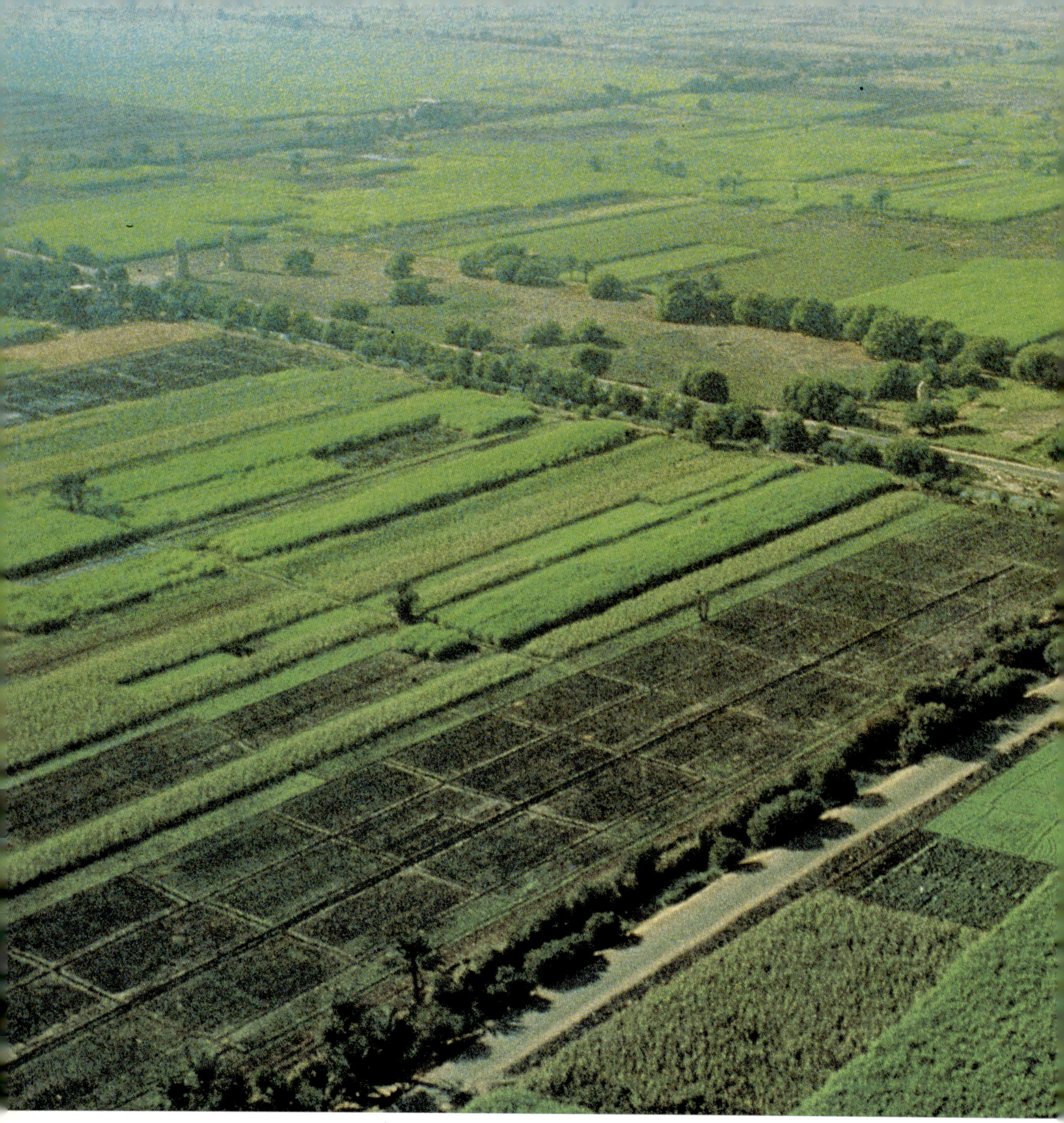

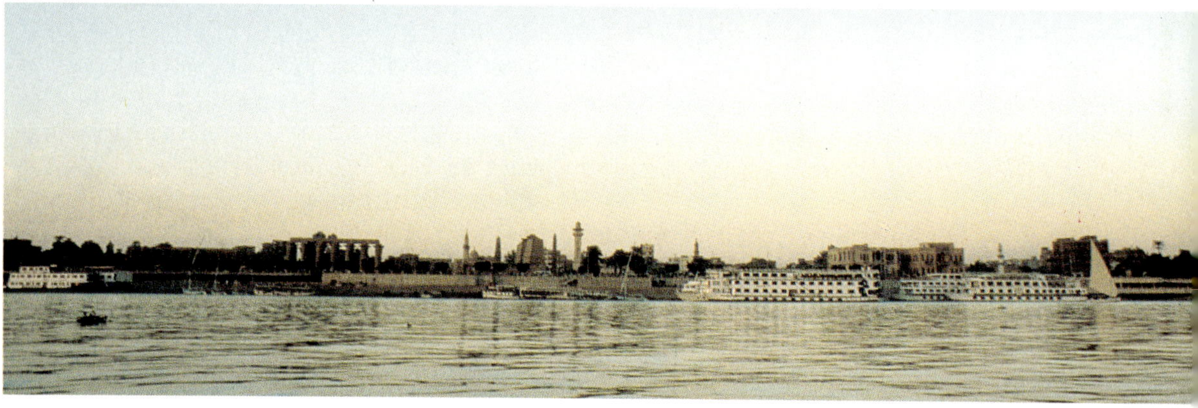

The river Nile.

tians of trying to restrain Ethiopia's development plans and refuse to agree to the water release demanded by the downstream states, saying that it makes their dam unviable. Neither the Egyptians, the Ethiopians nor the Sudanese are being particularly unreasonable from their own points of view.

6. The Sudanese say that the water resources of the White Nile are the sovereign assets of Sudan and that they must be allowed to do with them as they see fit within reason. Newspapers in Khartoum initiate a provocative campaign in which one editorial even suggests that if water is so precious to the Egyptians they might be charged for it. This is picked up by a popular paper in Khartoum with the headline "METER THE NILE!" It is reported in Cairo as if it is official Sudanese policy. Although the Sudanese delegation assure the Egyptians that no such scheme is even under discussion, tension mounts.

7. The Egyptians state that neither the Sudanese nor the Ethiopian positions are acceptable.

8. The Organisation of African Unity and the Arab League work behind the scenes to encourage the parties to negotiate a binding proposal for the shared management of the Nile but progress is slow.

9. The Egyptians, growing more concerned, appeal to the United Nations, seeking a Security Council mandate to allow the Secretary-General to organize an Inter-Governmental Conference to adjudicate the water resources in the region. Britain and the United States support and propose the Egyptian resolution, perhaps mindful of their debts to President Mubarak for his support in the Gulf War of 1991.

10. The Security Council passes a resolution to resolve the issue by diplomatic means, although China and the USSR both express reservations about the potential internationalization of the water affairs of sovereign nations.

11. The Sudanese and Ethiopians, however, lobby the Third World members of the General Assembly on the grounds that the Egyptians are in the pockets of the USA and its European allies, and they are acting as "imperialist stooges". France abstains in the Security Council vote and without notice, proposes an independent French plan for a regional conference.

The General Assembly refuses to endorse the Security Council's resolution, so the UN is divided.

Stalemate

12. The stalemate continues until 2002 when the Ethiopians complete their dam on the Blue Nile.

13. The Sudanese, with help from Germany, which had been involved in the original scheme, have now completed work on the Jonglei Canal and are more determined than ever that there should be a renegotiation in the allocation of water from the canal, which increases their allocation. They stress that the canal is essential for the development of their country. They restate their position about sovereign control over the Nile from the abortive 1999 Nile Summit and make it clear that they will act unilaterally if needs be. This could mean abrogation of the 1959 bilateral agreement.

Crisis

14. Global warming is affecting the North American crop belts, driving international grain prices to new heights. Loss of land in the Nile Delta has continued, despite some sea defence work. Egypt cannot afford to protect itself on the scale that the Dutch are doing. Egypt, which has seen a steady increase in the amount of staple food that it needs to import, suffers particularly

badly in a shrinking world grain market. There have been food riots in Cairo.

A stable Egypt continues to be seen as a shared policy goal by the Russian Federation, the USA and the USE (United States of Europe).

The domestic situation worsens, with the government looking increasingly vulnerable to its internal nationalist opposition. It recalls the robust stand of Zaghlul and of Nasser. It emphasizes that Egypt is the Nile and appeals to the most basic instincts of all Egyptians in the campaign slogan: " No Nile, No Egypt, No home".

So the government decides to divert attention from the domestic situation and plans decisive action to resolve the long-term security of Egypt's water supply.

15. The Egyptian government, still smarting from the débâcle at the UN, judges that a majority of the permanent members of the Security Council will privately support their attempts to protect the Nile water supply and steady the domestic political situation. They calculate that their allies would be anxious not to see a militantly nationalist or worse, fundamentalist Islamic government in Cairo.

16. Egyptian Special Forces, trained by the US and associated with the disgruntled Northerners in Sudan – provide covert assistance for a coup in Khartoum, and topple the Sudanese government. In return the rebels agree to be more accommodating about water allocation and agree to a binding treaty. The coup in Khartoum is successful but it leads to Southern insurrection. The Egyptians find themselves drawn, increasingly publicly, into a guerrilla war in the South as they are forced to protect their White Nile water supply by providing support for the new government.

17. Determined to ensure Blue Nile supply from Ethiopia the Egyptians issue an ultimatum to the Ethiopians that they must negotiate a new treaty over the Nile proposing an international body of all riparian states. The Ethiopians do not respond, underestimating the top priority which the Egyptians give to the issue.

18. The Egyptian Air Force flies a warning sortie over the dam, while commandos cripple the dam's turbines.

19. Shocked by the escalation, no longer able to count on Sudanese support, the isolated Ethiopian Federation agree to talks and a new treaty is eventually signed by all nine riparian states. The Italian government takes a leading role in arranging a substantial technical assistance programme and trade credits for Ethiopia which are offered as an inducement, and as a way of helping the Ethiopian Federation to save face.

19. The international community makes ritual protests about Egypt's actions but privately many countries are glad to see an end to the dispute.

Chapter 5:
The Best of Times

Were the doomsters wrong?

Does it have to be that way? When concerned scientists and policy makers began to become conscious of the demand that an exploding world population would place upon the natural resources of the earth, their first instinct was that the only way out was to consume less. This was the message promoted powerfully in *Limits to Growth* by Professor Dennis Meadows and his team. It was the first attempt to see how many superficially different things interacted with each other in a world simulation model. The supply of raw materials, food, energy and pollution were a few of the main categories which were taken and projected into the future under the overarching influence of a burgeoning world population. *Limits to Growth*'s model showed on repeated runs that the most likely outcome was crisis.

Limits to Growth was heavily criticized at the time of its publication both for the crudity of its model (which aggregated global trends and failed to make regional differentiation) and – unfairly – because attempts were made to take its modelling hypotheses and turn them into precise predictions of the moment at which oil or some other essential ingredient of industrial life would run out. When as a result of the oil-price shocks of the early 1970s the age of cheap energy ended never to return, it was with much relief that many politicians and commentators pointed to the new discoveries of oil in the North Sea and in other places, made possible by the sudden rise in the price of oil which made such exploration and exploitation financially viable. They saw this as a sign that the "doomsters", as they called Meadows and his team, had been wrong.

In the long term, of course, the doomsters are not wrong. The earth is finite and its raw materials will, without any change to current political attitudes, eventually be rationed, first by price as the

These composite images, taken from space, show millions of lights burning at night in North America (*top*) and Western Europe.

market mechanism comes into play and finally by exhaustion. But that moment is still far off even given the profligate use of raw materials which has been the mark of industrial society during stages of initial growth.

Energy is our guide

When we looked at the worst of times, we drew on case studies which covered a wide range of circumstances. To look at how things could be different, we will use a focused approach and concentrate on world energy production and use. There are four reasons for doing so.

■ The first is that energy consumption and use is the human activity which is most visibly, sometimes violently, at the interface between humankind and the earth. We dig or pump energy sources from the earth and throw huge amounts of heat and combustion pollutants straight back into the biosphere.

■ The second reason follows straight from the first. The contribution of fossil fuel energy use to the accumulation of greenhouse gases like CO_2 is direct. And the prospect of global warming leading to climate change is the central issue on the emerging agenda. This was made clear by the report of Working Group One on the scientific assessment of climate change in the UN IPCC process, which calculated from existing Global Circulation Models that under a "business-as-usual" scenario, we might expect a 1°C rise in mean global temperature by 2030 and of 3°C by 2090. The reports were presented at the Second World Climate Conference in Geneva. Its final statement, released on 7 November 1990, observed that:

> Climate issues reach far beyond atmospheric and oceanic sciences, affecting every aspect of life on this planet. The issues are increasingly pivotal in determining future environmental and economic well being. Variations of climate have profound effects on natural and managed systems, the economies of nations and the well being of people everywhere.

John Houghton, Director of the British Meteorological Office and Chairman of the Working Group, wrote two days later in the *London Financial Times* that:

> Never before in the history of science has there been a scientific problem that has become so dominant in the political arena and on no other subject have scientists, economists and politicians been forced together in such close dialogue.

Therefore, given the close coupling between energy use and the central problem of global warming, energy is the most efficient issue upon which to centre.

■ Given this, and also that energy production and use underlie so many other activities that we may take it as a lowest common denominator, a third reason is that if we can see how attitudes and practices may be changed in energy use, which is so basic an activity, we may feel some confidence that changing attitudes and practices in dependent activities may follow consequent on winning that larger battle.

■ Finally, as we shall see in Chapter 7, energy production and use is a field in which technologies developed for military purposes may have particularly powerful and immediate applications.

The first industrial revolution was fuelled by coal. Compared to the amount of coal probably existing and ultimately recoverable, the world has used only a small portion of its potential. About 150,000 quads (quadrillion, 10 to the fifteenth

British Thermal Units) remains which at 1988 rates of use could last 1500 years. But coal was displaced during and after the Second World War by oil as the primary fuel of advanced industrial societies. At the same time, world energy consumption rose enormously: from 60 quads in 1937 to 321 quads in 1988. Whereas coal had accounted for 74 per cent of the primary fuel in 1937, by 1988 it had fallen to 30 per cent, whereas oil and natural gas together account for nearly 60 per cent. Yet the world's supplies of natural gas and oil are far more limited than those of coal.

Oil is used overwhelmingly to fuel the modern transportation system whereas coal in the modern industrial economy goes principally to electricity generation and to sheer waste. Two-thirds – 33 million barrels of oil equivalent per day [m.b.o.e.d.] – of the 48 m.b.o.e.d primary energy burned to generate electricity is lost in conversion – about 35 per cent of the world's daily coal burn is simply lost. That's about the same as the total output of hydro-power and nuclear power worldwide, combined. Despite this spectacular wastefulness, coal remains plentiful. Yet those who dismissed the warnings of the authors of *Limits to Growth* are unwise to laugh too swiftly; for the most striking contrast between the situation for coal – the fuel of the first industrial revolution and oil – the fuel of the second industrial revolution – is that oil reserves *other than in the Middle East* have already been heavily depleted. The geopolitical significance of the fact that 2500 quads of the total remaining 7000 quads is located in the Middle East is stark.

Dependence upon cheap oil in the decades between 1940 and 1973 has left the industrial world with a pattern of energy consumption which cannot be described as stable or sustainable for more than the short term. So in this sense, the authors of *Limits to Growth* were right and it is a pity that their message was clouded by unwise attempts to speculate too precisely upon the time span of each of the raw materials remaining.

For fifteen years after the publication of *Limits to Growth*, the debate about the sustainability of our life style was very much cast in these terms. In so far as the poor were considered, it was to consider how aid policies could help give them access to fuels and finished products that were becoming steadily more expensive for them in real terms as the value of many of their primary products fell in comparison. The terms of trade moved sharply against the poor in 1973. In so far as the environmental consequences of loading the atmosphere with the products of burning these vast quantities of fossil fuel was concerned it was noted and disapproved of, but frankly not thought likely be a primary driver of change. There was no conclusive evidence.

The points of the triangle joining together environment, development and resources were not yet joined.

Two experts on energy for the developing world Amulya K. N. Reddy of the Indian Institute of Science in Bangalore and Jose Goldemberg, former President of the energy company of the state of São Paulo, and currently Secretary for Science and Technology in the Brazilian government have described this attitude to energy planning as "Consumption Obsessed, Supply Biased". We can call it COBSUB for short.

In 1987 there came a breakthrough in understanding. In many ways, it was the next major step on from the breakthrough that was represented by the *Limits to Growth* report. After four years' work, the World Commission on Environment and Development, the Brundtland Commission, reported. *Our Common Future* shows how the Commission began its work aware of the need to look for interconnections between the various crises some of which we described in the previous chapter. But stepping beyond the concerns of the early 1970s, the Commission sought to explain the interlocking between environment, development and resources.

> This Commission believes that people can build a future that is more prosperous, more just, and more secure. Our report, *Our Common Future*, is not a prediction of ever increasing environmental decay, poverty and hardship in an ever more polluted world among ever decreasing resources. We see instead the possibility for a new era of economic growth, one that must be based on policies that sustain and expand the environmental resource base. And we believe such growth to be absolutely essential to relieve the great poverty which is deepening in much of the developing world. (From *One Earth to One World*, an Overview by the World Commission on Environment and Development, *Our Common Future*).

The vital concept which the Commission placed at the centre of its thinking was that of *sustainable development*. By sustainable development, the Commission meant finding new paths of economic and social development that, "meet the needs of the present without compromising the ability of future generations to meet their own needs".

Defined like this, sustainability is not a new concept. On the contrary, it is very old. During the last generation, anthropologists have greatly increased our knowledge of the amazing and varied ways in which so called "primitive" peoples managed to live in harmony with difficult environments where modern "civilized" peoples – city dwellers with soft hands and soft bellies – could never survive. Marjorie Shostak has given us a privileged insight into the life of one such community through her biography of Nisa, an !Kung Bushman woman from the Kalahari Desert of Southern Africa. We have similar studies of the skills of aboriginal communities in Australia and of the indigenous peoples of Brazil.

While images of hunter-gatherer peoples like these may be the first to spring to our minds, we should remember that much of the history of Africa, for many thousands of years until the late nineteenth century, was of communities not only of hunters and gatherers but also of farmers who found ways to adapt an intense and minute knowledge of their local environment, its plants, animals, soil and climate, to agricultural systems of great elegance, developed by trial and error, which enabled people to live sustainably with the most environmentally difficult continent on the planet. Finding ways to live and live well in Africa from the Sahara Desert to the Cape of Good Hope has been one of the greatest achievements of African civilizations and forms one of the finest chapters of the human story written by Africans.

A tragic irony has been that this achievement has been invisible to the eyes of the conquering peoples of the last half millennium. To European eyes, the chaotic intermixture of different plants to be found in a field in Sierra Leone was a sign of the backwardness and incompetence of West African farmers; it bore no resemblance to the neatly trimmed hedges and manicured pastures of Kent. Yet, as the agricultural historian Paul Richards has revealed, we can see that, far from chaos or ignorance, such interplanting of crops lay at the heart of the success of Sierra Leonese peasant agriculture. Quick growing and broad leafed plants could shade and protect from rain and sun slower growing and tender shoots. The mounds of earth apparently carelessly cast up around the base of plants were constructed at an angle such that torrential rain would fall off, protecting the young plant in a rainstorm, and yet not at so steep an angle that the plant itself was suffocated.

All these were insights that arose not from science but from experience. All these were essential to creating

An Amazonian Indian.

sustainability in tropical agriculture. All these were despised by ignorance and – one of the great tragedies of our century – the owners of this knowledge were taught to despise or belittle what they knew and to value the products of "coca-colonization" more highly. In this way they were doubly robbed, both of the principles of sustainability in their small communities and of their ability to recreate it should they escape the spell of the conqueror.

So we can grasp the concept of sustainable development most easily in small scale societies and in agriculture. It is easier to see the further away we get from Europeanization. Yet Europeanization has been the dominant feature of the last five hundred years.

As we explained in Chapter 3, powerlessness, the sense of helplessness in the face of forces which seem irresistible, leading to denial can easily become a common response to the new threats which bear down upon us. As we in the industrial world experience these feelings, we share a little the sense of powerlessness and hopelessness which has been the lot of many of the non-European peoples of the world during the period of European ascendancy. Many of them did indeed descend into alcoholism and despair; but, amazingly and encouragingly, once we know to look for it, we can begin to find evidence of remarkable histories of resilience and adaptation to complement the better known but atypical episodes of non-European revolt against colonial rule.

People found ways to adapt and to live under vastly changed circumstances. Their example stands before us as now, on a global scale, we have to find the political will to create sustainability at the largest scale as well as at the smallest. If we do nothing, if we allow denial to corrode our self confidence and to ignore the problems, then the worst of times may happen. Yet, as we argued in Chapter 3, if we can assess the new threats to our security and apply to that analysis the medical model and our well-tried methods of military threat assessment, we may find ways to generate the political will to try to connect the corners of the Brundtland Commission's triangle of environment, development and resources.

If we do so, have we the resources to hand which would enable us to make choices that take us away from the worst outcomes?

Here there is much to be optimistic about. In the rest of this chapter and in Chapter 7 we shall review some of the new means to hand, many very new indeed and in a state of constant and fast development, which, if we wish, we can turn to this task in the field of energy production and use. But the prerequisite is to share President Carter's sentiment, expressed at the time when he advocated a massive programme of research into energy conservation in the United States that such action is "the moral equivalent of war".

Towards a global benchmark for sustainability

The Brundtland Commission identified two key concepts within the idea of sustainable development. The concept of *needs*, and especially the needs of the world's poor to which overriding priority should, in the Commission's view, be given, and the idea of *limitation*: limitations imposed by the state of technology and social organization on the environment's ability to meet present and future needs. Four years on we can see that this is, perhaps, the first stage of definition of the concept of sustainability. The problem over the last few years has been that as the term has become more widely used, its edges have not become less blurred. In *Blueprint for a*

Green Economy, David Pearce and his colleagues include a Gallery of Definitions of Sustainable Development. The need, then, is to find the lowest common denominator which would enable us to define sustainability in a sharp sense. Amulya Reddy and Jose Goldemberg have given us a boost towards this end with their particular historical and comparative treatment of the concept of energy intensity.

The concept of energy intensity has much to recommend it. It is easy to understand. Energy intensity expresses the amount of energy (in equivalent metric tonnes of petroleum) needed to produce a given unit of gross domestic product. The second attraction is that it is thus universal: this function can be tracked for different types of societies at different times. So a third attraction is that, as used by Reddy and Goldemberg, it permits historical comparison between the developed and developing countries. What does it show?

The first and most striking feature is that the energy intensity function of countries has the same shape. During the period of initial industrialization the amount of energy consumed (to create infrastructure, for example) increases sharply and then a "hump" is passed after which energy per thousand dollars of GDP falls over time. Secondly, countries starting later in the process seem able to cross the "hump" at progressively reduced levels of energy consumption.

Energy intensity does *not* measure sustainable development. What it does do is to provide a clear indicator of the historical trends of energy use and, since energy use underlies the whole spectrum of economic activity, reduction of energy intensity can become a politically realistic component of a strategy towards sustainable development. It is where mankind's demands upon the biosphere are most direct. So it is a concrete way of framing "needs". The other component of sustainability is to do with the switch from wasting to renewable resources in producing such energy as is required to fulfil needs. Of that more below.

What is it that gets countries across the "hump"? What is it that produces the striking pattern and reduction of the maxima in country after country, and if that pattern is to be followed what is it that can give us the hope that developing countries can manage the transition across their hump in due course at a lower intensity of energy use? In all cases the answer is obvious and it is the improvement in energy efficiency and in technology. It is here that tremendous developments are in train at present. But crossing the "hump" has two other prerequisites as well.

New technologies cost money, especially if they are adopted at a point in the life span of existing industrial plant when it is not yet worn out. And people require both know-how and the ability to control the technology, rather than to be controlled by it. Reddy and Goldemberg express this new objective by the acronym DEFENDUS: Development Focused, End-Use Oriented, Service Directed. DEFENDUS planning of development stands in contrast to COBSUB (Consumption Obsessed, Supply Biased) approaches. What DEFENDUS does is to incorporate conservation and the favouring of renewable sources under the "needs driven" criterion of sustainable development proposed by the Brundtland Commission report. It is strong on what David Pearce and his Green Blueprint colleagues call "intergenerational equity". In other words, being kind to your unborn grandchildren.

Four stages in energy history

On the road to sustainable development we have to change both the intensity of our energy use and also the source of our

Source: "Energy for Planet Earth" by Ged R. Davis. Copyright © 1990. Scientific American, Inc. All Rights Reserved.

CRUDE OIL: 59   COAL: 41   NATURAL GAS: 29   HYDRO-POWER: 9   NUCLEAR POWER: 6   OTHER: 22 (INCLUDES TRADITIONAL FUELS)

REFINERY USE/LOSS: 4
NONENERGY USE: 11 — 8 | 1 | 2
TRANSMISSION LOSS: 3

INPUT: 6 | 20 | 9 | 7 | 6
ELECTRICITY GENERATION: 48
OUTPUT — CONVERSION LOSS: 33 | 15
TRANSMISSION LOSS: 2

TRANSPORTATION: 27 — 26 | 1
INDUSTRY: 42 — 7 | 13 | 8 | 7 | 7
RESIDENTIAL AND COMMERCIAL: 43 — 8 | 6 | 8 | 6 | 15

MILLIONS OF BARRELS OF OIL EQUIVALENT PER DAY

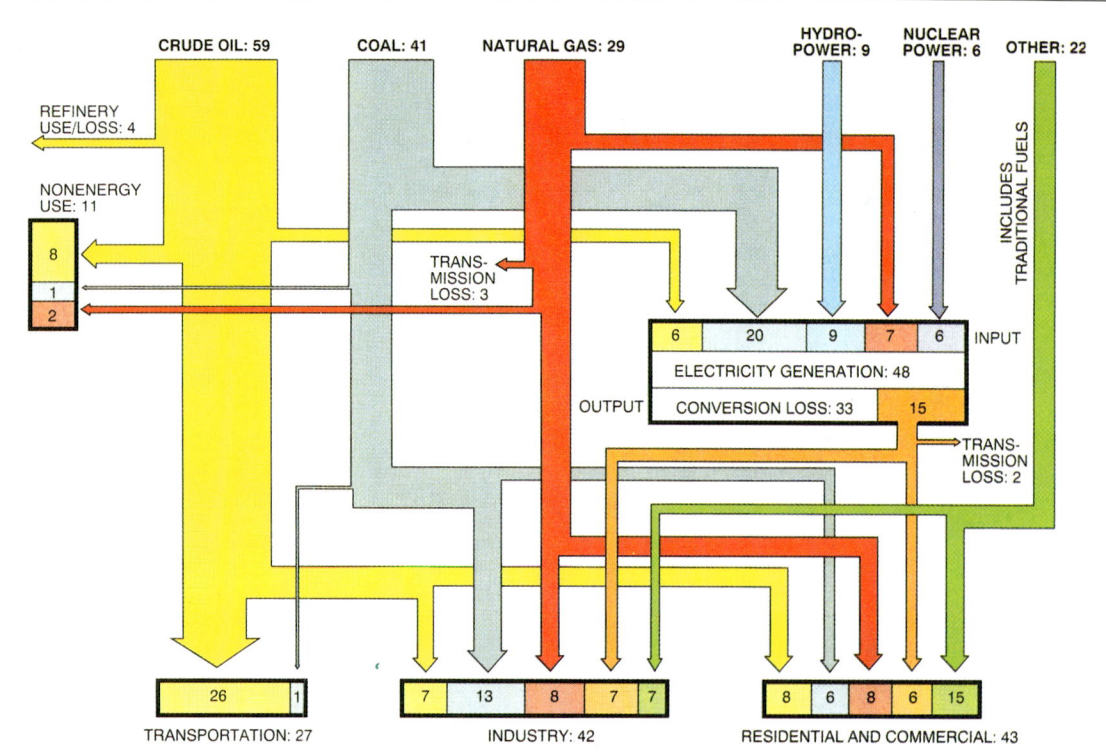

World energy flows for 1985 show that fossil fuels are highly versatile. Crude oil (yellow) is converted to gasoline, diesel and aviation fuel for transportation. Both oil and natural gas (red) are widely used by industry and by residential and commercial consumers, but only a small percentage of each generates electricity. Most of the world's coal (light blue) powers industry or generates electricity. Hydropower, nuclear power and other energy sources (biomass, solar energy and wind power) make up 22 per cent of today's primary energy supply.

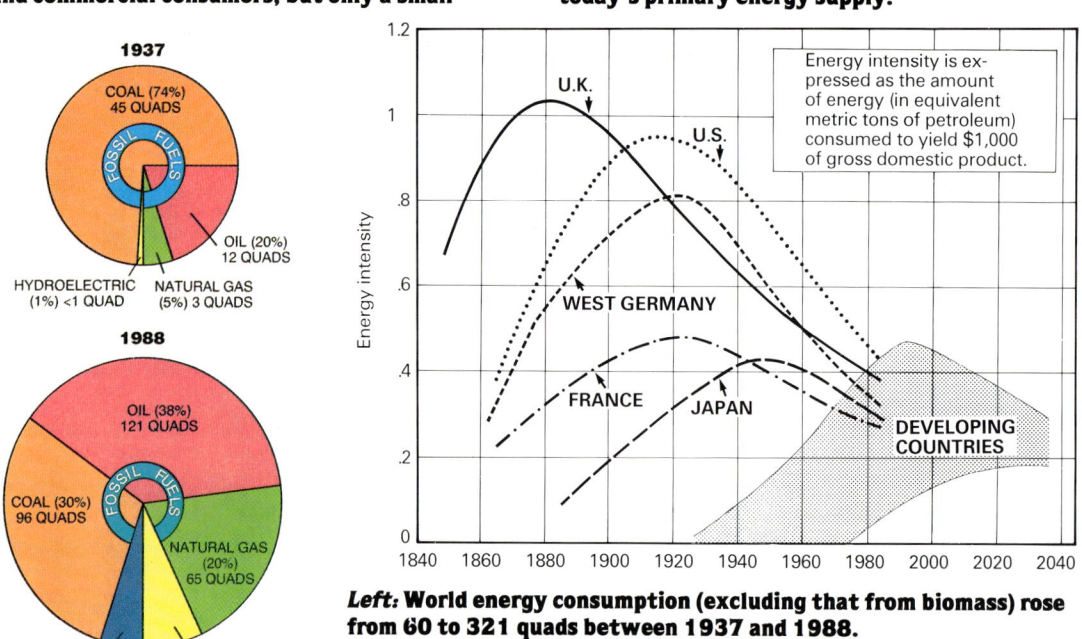

1937
- COAL (74%) 45 QUADS
- OIL (20%) 12 QUADS
- NATURAL GAS (5%) 3 QUADS
- HYDROELECTRIC (1%) <1 QUAD
- FOSSIL FUELS

1988
- OIL (38%) 121 QUADS
- COAL (30%) 96 QUADS
- NATURAL GAS (20%) 65 QUADS
- NUCLEAR (5%) 17 QUADS
- HYDROELECTRIC (7%) 22 QUADS
- FOSSIL FUELS

Energy intensity is expressed as the amount of energy (in equivalent metric tons of petroleum) consumed to yield $1,000 of gross domestic product.

Energy intensity curves: U.K., U.S., WEST GERMANY, FRANCE, JAPAN, DEVELOPING COUNTRIES (1840–2040)

Left: World energy consumption (excluding that from biomass) rose from 60 to 321 quads between 1937 and 1988.
Above: In industrialized countries the energy intensity (ratio of energy consumption to gross domestic product) rose, then fell.

"Energy from Fossil Fuels" by William Fulkerson, Roddie R. Judkins and Manoij K. Sanghvi. Copyright © 1990. Scientific American, Inc. All Rights Reserved.

primary energy supply. These choices we will make most reliably if we use the DEFENDUS criteria. Under these criteria the challenge now is to effect the switch from wasting to renewable energy sources. Daunting as this is, we have at the same time to effect a second transition which would logically precede the switch from wasting to renewable resources. This is a transition from the era of cheap energy to that of costly energy. These major transformations in the underlying energy regime of the world will be the third and fourth respectively to have occurred since industrialization began.

Stage One	Stage Two	Stage Three	Stage Four
Wood to Coal	Coal to Cheap Oil	Cheap to Costly Oil	Fossil to Sustainable

At the beginning of this chapter we illustrated the second shift, that from coal to cheap oil. It shows how within the span of a generation, the advanced industrial economies made a reorientation of great scale and comprehensiveness. So these things can be done. But there was an important difference in the intellectual force which drove the first great transformation from wood to coal (renewable to wasting asset) and that which underlay the transition from coal to oil.

This difference lay at the heart of the bitter complaint which the engineer-turned-historian, L. T. C. Rolt, made against the late nineteenth-century transition from "engineer-led" to "science-led" technological development. In *Victorian Engineering*, Rolt framed his view of the century around his attack upon Adam Smith and the principle of the division of labour. As an engineer, Rolt knew that without the division of labour, the technological advances that made Britain the workshop of the world would have been impossible. He rightly credited Henry Maudslay as one of the greatest of the early mechanics and pioneers of workshop precision.

Maudslay died in 1831, but his influence on British engineering lay upon the rest of the century, for he it was who first realized the importance of precision, and his first screw-cutting lathe was the ancestor of modern lathes because it had the property of being able to be self propagating. Accurate and replicable machinery was essential to permit the specialization which made possible the intense development of knowledge that came from the division of labour. The addition of scientific to trial and error methods greatly accelerated this intensity and this development. Yet Rolt rebelled against it. When he contrasted his heroes, the great road and canal engineer Thomas Telford or the railway engineer Isambard Kingdom Brunel, with their late nineteenth century successors, he saw lesser men in the later generation because they lacked both the all-encompassing knowledge, the mixture of the practical and the technical and, in Rolt's view, a sense of social responsibility about their products which had been the mark of the early engineers.

He felt that this integrating sense of technology in society had been lost with tragic consequence. When, at the end of his book, he looked to the future, he offered a prescription which still serves us well.

> . . . we need to recapture the mood of optimistic zest and eager inquiry that animated the pioneers in the last decades of the eighteenth century. But whereas they directed their energies to the pursuit of knowledge and invention we need to devote ours to breaking down the barriers between the isolated compartments of knowledge. The urgent need is for synthesis no matter how difficult such a goal may be . . . until with humility we strive for this creative synthesis, not all the millions that are now spent on research can recover lost certitude and reanimate the sprit of the pioneers. Nor can individual man achieve a like stature, or feel like his forefathers that he is wholly responsible for his destiny. (L. T. C. Rolt, *Victorian Engineering*)

The DEFENDUS criteria tell us that we must find a way to reintroduce the sense of purpose and direction and spirit of the pioneering engineers while at the same time drawing upon all the benefits that must come from the division of labour. Professor Timothy O'Riordan of the School of Environmental Sciences at the University of East Anglia, addressing a group of environmental activists on the north Norfolk coast at a meeting to protest about the dumping of untreated sewage effluent into the North Sea, distinguished for them between *paternalistic science* done by "men in white coats in laboratories who pontificated on the basis of expertise" and *precautionary science*, where "you act in advance of scientific certainty ... you ought to give yourself leeway, or buffering protection against ignorance." Precautionary science, he said, harmonizes with *participatory science* – with space for people who "believe very firmly about what is right and wrong about pollution and all the things that you hold dear".

So the lesson of the transition from coal to cheap oil is only that it was possible to do it and not at all about the motive force which led to that change. For the motive force at that time was *dissociative* and driven by other imperatives than those of sustainable development. In fact the midwife of the petrol engine and of the age of oil dominated industry and transport was in both cases war. The First World War hugely stimulated the development of the internal combustion engine, which became in the interwar years a plaything for the rich in all countries (except the United States where Henry Ford's Model T made the motor car accessible to ordinary people earlier than elsewhere) and to an increasing but nowhere dominant degree a challenge to the railways as a prime mover of freight. The Second World War completed the ascendancy of motorized land transport and also ushered in the age of mass air transport.

If we can find the political will to make investment decisions in primary energy fields development focused and end use oriented, the technological possibilities for moving forward are bright indeed. Here are two detailed examples. One comes from the advanced industrial world and shows how the type three transition from cheap oil to costly energy can be very quickly made. The other example comes from the developing world and shows how in the late twentieth century it really is possible for the poor and backward to leapfrog a whole stage of industrial development into a sustainable future and to accomplish the transition from wasting to renewable energy at stage four.

Enter the negawatt

It is possible to make the first moves in the shift from a regime of cheap energy to one of costly energy without the accompanying conceptual shift which the DEFENDUS criteria represent. Quite simply, it's good business to make more goods with less input; and if the inevitably short-term logic of the market perceives a shortage of raw materials which is reflected in an increase in price the capital investment in technology which reduces energy intensity will be made. That is the logic which sent the oil companies into the North Sea following the rise in oil prices in the early 1970s. In some industrial economies, notably in Japan which lacks any substantial fossil fuel resources, substantial investment in more energy-efficient plant meant that Japan's electric use per tonne in major industries such as paper and steel and car making, began to fall ahead of reduction and improvements in energy intensity in the United States or in Europe. California reduced its electric intensity by 18 per cent from 1977-86, which was a marked improvement on any of the other states in the union.

The response to reduced use and to the general slowdown in the advanced market economies which followed the oil price shocks was to pull the oil price back down again. An important question for the 1990s is what the long-term consequences of this period of depressed oil prices will be. The American energy expert Professor John P. Holdren believes that with the fact that most of the cheapest oil and gas (except for the huge pool of oil under the Middle East) has gone, and with the absolute increase in energy demand which comes with the explosion of world population since 1960 (from 3 to 5.2 billion) those who have seen the fall of oil prices during the 1980s as a sign of a return to the old regime are deluding themselves.

> There are . . . a variety of other energy resources which are more abundant than oil and gas. Coal, solar energy and fission and fusion fuels are the most important ones but they all require elaborate and expensive transformation into electricity or liquid fuels in order to meet society's needs. None has very good prospects for delivering large quantities of fuel at costs comparable to those of oil or gas prior to 1973 or large quantities of electricity at costs comparable to those of the cheap coal fired and hydro power plants of the 1960s. It appears then, that expensive energy is a permanent condition even without allowing for its environmental costs. (J. P. Holdren, Professor of Energy and Resources, University of California, Berkeley)

In addition to increasing costs from scarcity, Holdren suggests that energy costs will be increased in real terms as the shift away from Consumption-Obsessed-to End-Use-defined views of energy marches swiftly forward. The increase will be by what he calls "internalized costs" which mean the application to the price of energy tariffs designed like carbon taxes to shift fuel preferences in a more environmentally friendly direction, thus reducing what Holdren calls the "external costs" which are the consequences for the biosphere of environmental disruption as a consequence of energy use.

The principle of intervening in the energy market in this way is one which was strongly endorsed by the World Commission on Environment and Development. It saw that only by making the third corner – the environment corner – of the triangle of development, environment and resources visible and active in the operation of the international economy, would effective action be taken: action both *through* market forces and *beyond* market forces by environmentally informed intervention.

Some environmentalists have argued that the conventional practice of discounting in investment decisions (writing off the value of assets over a period of time as they are presumed to wear out or to become obsolete) works against the environment. The argument has been made that high discount rates tend to increase the temptation to maximize the use of exhaustible materials and to cause investment decisions to be taken with little concern for the future because, as John Maynard Keynes observed, "In the long run we are all dead" and the point of economic activity in the conventional sense is to maximize gratification for the present in the shortest space of time available. However, David Pearce and his colleagues in *Blueprint for a Green Economy* do not agree with this argument. They conclude that interference with the discount rate is *unlikely* to favour environmental concerns and believe that intervention should be conducted at a far more fundamental level – that of structuring the relative values of commodities under DEFENDUS criteria.

A splendid illustration of why high discount rates, favouring investments which make a quick return, can work in the interests of the environment comes from a revolution in electrical end use technol-

The Three Mile Island accident in 1979 was effectively the death-blow to civil nuclear power in the USA. No new plants have been ordered since then. After the explosion at Chernobyl in the Ukraine in 1986 (inset), the nuclear plant was encased in a concrete sarcophagus. The cloud of radiation that spread out from the exploson crossed vast tracts of Europe, polluting water and food.

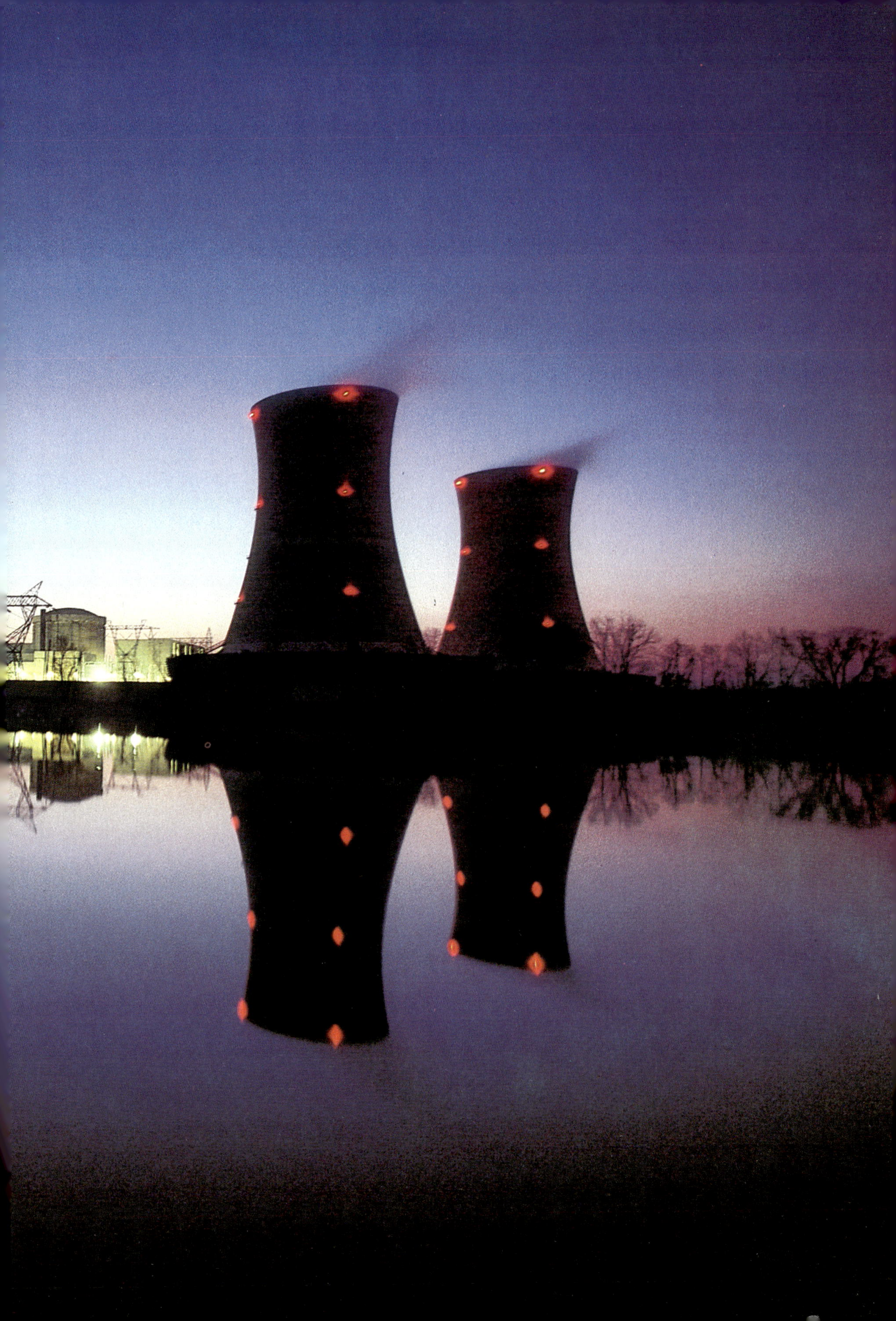

ogies which is now in train.

The key technology of the present industrial revolution through which we are now living is the revolution in the management of information. In the new Stone Age, in the land of the two rivers, the Tigris and the Euphrates, occurred the last great revolution in information technology when mankind acquired codified systematic literacy. This made possible a revolution in the way in which we relate to information: instead of having to hold everything that we know inside our heads, literacy made it possible to write down information in a physical form, to store it away and at a later stage, having forgotten it, to retrieve it, to read it, to remember it and then to reconsider it in a different way to that which is possible in an oral culture.

Our ability to store and retrieve information in a written form was hugely increased by the print revolution of early modern Europe. Gutenberg and his successors, through printing, made it possible to multiply and distribute messages and until recently we have tended to think of this revolution as being equal in scale to that of the acquisition of initial literacy. But living through the computer-driven information technology revolution of the late twentieth century, we know that this was wrong.

Printing greatly increased our ability to "store and retrieve" information; it did not greatly increase our ability to *manipulate and process* information. That is the thing that literacy gave at the beginning of civilization and it is in this area that the modern revolution in information management and in telecommunications is having its effect.

In 1960, in his ground-breaking book *Understanding Media: The Extensions of Man*, the Canadian thinker Marshall McLuhan predicted at the very beginning of the computer age that the revolution in information management and in telecommunications would substitute electrical journeys down wires on the airwaves for oil-driven journeys on wheels. In fact, we already see the beginnings of this with the advent of "telecommuting".

The point is, of course, that electricity is the primary energy that drives these leading-edge technologies of the future and it has been common for "pro-electric-use" bodies like the American Electric Power Research Institute to argue that the rise in electric use within the energy mix, closely coupled to the rise of gross domestic product was both desirable and encouraging. Yet as the Resource Policy Research Center at the Rocky Mountain Institute in Colorado suggests, it is not at all clear that increasing wellbeing and wealth necessarily involve a continuing rise in electricity consumption. We recall the sobering fact that two-thirds of the calories burned in indirect electricity generation are thrown away. So electricity is not a *source*, but a very costly, high grade *product* of energy use. Something which can run a computer should not be squandered by heating a bar fire (the most environmentally costly means of space heating).

In contrast to the pro-electric use groups, the Institute has calculated that beyond the impact of electronics and telecommunications, there are other areas of electrical use where potentially enormous savings can be made as a result of the use of advanced technologies. The Institute has studied electricity use in the American economy and estimates that in the past five years the potential to save electricity has *doubled*, whereas the average cost of *saving* a kilowatt hour of electricity has *fallen by about two thirds*. At the same time the cost of installing new conventionally powered generating capacity in large "base load" stations has continued to rise steeply. Interestingly, it has also found that most of the best efficient technologies for energy saving are less than one year old.

The Rocky Mountain Institute's study of electricity saving potentials in the United States shows what could happen if decisions on the electrical energy cycle were driven by end-use criteria rather than generating-capacity criteria. In the United States, lighting consumes about a quarter of electricity. Converting to the best hardware available today could save a stunning *80 to 90 per cent* of the electricity used. Compact fluorescent lights, for example, use 75 to 85 per cent less electricity than incandescent bulbs and they last a lot longer.

The Rocky Mountain Institute calculates that retro-fitting with high-tech lighting throughout the United States would not only save between a seventh and a fifth of all the electricity used in the United States but would also improve productivity because it provides a better working environment. Writing about these dramatic new findings, three leading American energy experts, Arnold Fickett, Clerk Gellings and Amory Lovins say "This is not a free lunch; it is a lunch you are paid to eat . . . lighting innovations may be the biggest gold mine in the entire economy."

After lighting, improved control of electric motors by electronic adjustable speed drives and advances in refrigeration and freezer technology offer further tremendous savings with "pay back times" on investment that are very attractive to the business community, ranging from six months to three years. The best refrigerator technology can now consume 80 to 90 per cent less electricity than conventional models, televisions 75 per cent, photocopiers 90 per cent and computers 95 per cent. The Rocky Mountain Institute put its money where its mouth was and, by installing super-efficient lights and appliances in its 4,000 square foot headquarters, decreased electricity consumption by a factor of ten, and eliminated *99 per cent* of the energy needed to heat space and water. The building is so well insulated that even though located in the Colorado Mountains it needs no furnace; and this new technology paid for itself within a year.

Nor do the potentials for savings which would enable the advanced industrial world to switch from a cheap energy to an expensive energy regime end only at the point of use. Building electrical generating capacity is extremely expensive and, under high discount rates, has become increasingly unattractive to electrical utility companies. Furthermore, certain types of electrical generating plant have been among the first industrial enterprises to attract what Professor Holdren calls "internalized costs". The best example of this has been the worldwide collapse in the construction of nuclear power plants particularly under free market conditions.

Following the Three Mile Island accident in the United States in 1979, no new nuclear plants have been constructed and several extremely expensive schemes were abandoned during the process of construction. In 1981 five reactors were "mothballed" indefinitely in Washington State in the American Pacific Northwest, involving one of the biggest corporate write-offs ever. But it was calculated that it was cheaper to abandon than to complete the stations. The reason for this was that public opinion engaged in Professor O'Riordan's "participatory science" and forced a costing of the full nuclear cycle – construction of the plants, running of the plants, management of the waste fuel and eventual decommissioning of the plants. Furthermore, licensing hearings compelled the utilities to take account of these costs.

In the more open market conditions of the United States, private electrical utility companies quickly did their sums and stopped constructing nuclear plants. Even in the more benign environment of the United Kingdom, where it was government policy to try and encourage and protect nuclear power generation, it proved in the end to be impossible. The "privatization" of the British power-generating industry dealt the final blow as market logic ruled against such vastly expensive construction with so many unquantifiable and potentially huge ancillary costs and such a long payback time.

The United Kingdom remains very backward in the move to the next stage, driven by simple market logic, which is to make the *non-generation* of electricity as much a commodity to be traded as a unit of electricity. American utilities calculated that it actually paid them to help their consumers *reduce* their electricity consumption by giving them advice or even materials to better insulate their houses and to use more energy-efficient appliances rather than to pay the borrowing charges and construction charges of making expensive new electrical plant. Southern California Edison, for example, has given away more than 800,000 compact fluorescent lamps.

Such an example by the utility was complemented in the 1980s by the legislation in California which increased substantially the standards required for appliances and for buildings. These produced savings of 8.6 per cent on peak demand for Southern California Edison and cost the utility only about 1 per cent as much as building and running a new power station. This is another free lunch which you are paid to eat. New regulations proposed at the federal level and already accepted in principle would remove an electricity utility's disincentive to invest in efficiency by in effect compensating the utilities for the revenue which they otherwise lose by selling less electricity.

It means that electrical producers in advanced industrial countries could, with a bit more encouragement and popular pressure, be persuaded to re-examine their purpose. Are they there just to generate and sell the largest amount of electricity or are they there to provide in the most competitive manner the different end uses to which electricity is put? If the latter, then a market not only for megawatts of generated electricity but for *nega*watts of saved electricity may exist. Indeed, Fickett, Gellings and Lovins believe that making markets in negawatts maximizes competition to achieve savings and thus drives down the cost of increasing efficiency even faster and further.

A revolution in electrical technologies shows how advanced technologies, conventional market strategy and the new imperatives of environmental security can be harnessed together powerfully to effect the transition from a cheap-energy to a costly-energy regime which will reduce further the energy intensity of industrial societies and increase their quality of living.

But doing more with less energy is only part of the problem for the industrial world. Professor Holdren states starkly that none of the conventional alternative fuels to oil and natural gas looks likely to be able to substitute in a sustainable manner. So are there any other options? In the northern hemisphere the potential for capturing directly the power of the sun through photovoltaic cells is less than in the tropics. This is logically a technology of greatest applicability in the areas of highest sun fall, around the equator. Rather it will be the sea and the trees and the wind in the trees that will offer the greatest potential in northern latitudes and, even then, only to those countries which are geographically well situated to capture wind and waves or grow trees. For the northern hemisphere it may well be that in the short term the energy mix will have to give priority in the use of fossil fuels to those areas, for example in Central Europe, which are ill placed to exploit wind and wave power (although well placed to exploit tree power in the medium term). Those places which *can* do so should move quickly to the "type-four" transition in fuel sources, leaving the residual fossil-fuel burn to be done by those who cannot move as fast.

The United Kingdom is well placed, on the edge of the Atlantic Ocean and with strong prevailing winds, to exploit new energy sources for direct transformation of the power of nature into electricity. (Direct generation means that there is no two thirds conversion loss, as happens with thermal sources.) Wind power offers the most immediately available and worked-out example of how this may be done.

In path-breaking research, Michael Grubb of the Energy Research Group at the Royal Institute of International Affairs, Chatham House and formerly of the Cavendish Laboratory at Cambridge has calculated that offshore windmills have the potential to provide a very substantial amount of the *base-load* electricity required by the United Kingdom.

At first sight, this appears paradoxical. Base-load electricity is that which is required by consumers to be available all the time and, until Michael Grubb's work, it had always been presumed that an intermittent source, such as is a windmill upon which the wind may or may not blow at any particular moment, could never provide base-load electricity without the provision of extremely expensive and inefficient storage, for example in batteries. The costs of such storage, and the losses of power that would result from putting energy into and extracting energy from it, were presumed to far outweigh the fact that the wind is free, thus making the costs of power lost through generation and transmission in the production of electricity from fossil fuels look modest. What Grubb has shown is that this simply is not so.

At the heart of his argument was the recognition that advances in information processing and in switching of electrical distribution systems meant that it became possible to use a robust and modern grid system to take power from a *distributive network* of intermittently operating windmills and, by using computer power to match demand and available supply, to break the paradox and use intermittent sources to provide base-load power without the need for storage. Power-system analysts like Grubb are saying that for too long we have believed a technological myth: that a variable energy source comprising many small dispersed units cannot be compared with a large central base-load power station. What is increasingly clear is that a substantial portion – Grubb has suggested up to about 40 per cent – of total UK power could be provided from such variable sources without risk of power failure.

Thus, what energizes this new insight is not only a recognition of the transforming potential that the application of the computer revolution can have by helping to overcome the need for storage, but also a new way of looking at electricity: seeing it from the point of generation to the final point of use as part of a total system. In this, the provision of many different types of sources is an added advantage which has always been argued by electricity generators in industrial countries. But their argument has usually been in favour of a mix of nuclear, coal, oil and hydro power. What is now proposed is that wind power in the first instance must be brought into any calculation of a total energy budget in such places. Nor need it stop there.

In the case of the United Kingdom, another power source is available for exploitation. It beats upon the western coasts of the islands unremittingly. The power of the waves is there to be harnessed and by the mid-1990s, the UK could have had a gigawatt of wave-power capacity. But it won't.

In the late 1970s, Professor Stephen Salter of the University of Edinburgh designed a remarkable structure of steel and concrete, weighing 300 tons, just buoyant, which could lie in the water, oscillate with the power of the waves and turn this energy into compressed air which would drive turbines and generate electricity. The contraption had the shape of a duck's rump and for obvious reasons was christened "Salter's Duck". A scale model was built and demonstrated on Loch Ness. Waves hit it, smooth water behind. It was elegant, robust, met criteria of environmental security – and it worked. Calculations showed a likely "full cycle" cost of 5.5 pence per kilowatt hour. In June 1982, at a Wave Energy Conference at Trondheim, Norway, a figure of 3p/kWh was suggested by Clive Grove-Palmer, the engineer put in charge of assessing the Duck for the British Department of Energy.

Because of its position and former strength in marine engineering, Britain was in 1982 selected as "lead nation" for research by the EEC. But it was Norway that in 1985 completed the first full scale pilot plant (which produced electricity for 4p/kWh) and in that same year, the British government cancelled all future work on wave power. In 1987 Professor Salter's team dispersed. How could this have happened, and why?

Not long after his Trondheim paper, the Department of Energy's Advisory Council on Research and Development held a meeting on 19 March 1982 from which Grove-Palmer was excluded. The outcome of this meeting was a secret report which recommended stopping all British wave-power research to "save" £3 million at a time when £200 million was being spent on nuclear research and the government and nuclear lobby were pushing hard to build the Sizewell B nuclear power station. If wave power could be that cheap, it would undermine the nuclear programme. That

How they sank Salter's Duck

■ Shortly after the nuclear enthusiasts who ran the Advisory Council on Research and Development recommended sinking wave power, Clive Grove-Palmer, the sympathetic engineer, unexpectedly took early retirement and was replaced by a nuclear enthusiast with the task of running down all wave research.

■ Assessment of wave power was given to the Energy Technology Support Unit, based at the Atomic Energy Research Establishment at Harwell. This is like asking Dracula to look after a blood bank. The report of Gordon Senior, a consultant engaged to assess the Duck, had his conclusions altered and in some cases reversed without his knowledge. When he discovered this, as Mr Senior told a House of Lords Committee inquiry, "I objected and asked for my views to be made known to the Deptartment of Energy but was told that this could not be done and that I was bound by client confidentiality not to reveal my disagreement. I was also advised not to have further contact with the device team..."

■ In order to fiddle the figures, the Department of Energy had to introduce absurd capital costs for construction. The realistic cost was £208/tonne for the steel and concrete construction. The Department of Energy used £10,000/tonne. It also had to postulate absurd unreliability in the underwater power transmission cables: Norwegian figures suggest one failure per kilometre each 625 years. Actual experience in Orkney gave 333 kilometre years, but the Department of Energy said one failure per kilometre *per month* for the Duck...! (Later said to be a "misprint" for 10 kilometre years.) It also had to postulate very low availability; whereas waves on the Atlantic coast are there on average 40 per cent of the time and 70 per cent in winter, when electricity demand is highest. Much higher availability than from nuclear power.

was the reason that it had to be stopped.

The evidence became fully public in 1990 and demonstrates that British wave power was sunk as a result of a deliberate policy of dirty tricks and lying, with the objective of making wave power look more expensive than nuclear power. Some of the worst aspects are worth recording in detail because they are a case study of how far people will go to exercise denial if, as was the case for the nuclear power enthusiasts who dominated the British energy industry and the government bodies supposed to regulate it in the 1980s, their worldview was threatened.

The subversion of Salter's Duck in the United Kingdom is a tragic example of how far we still have to go in making the transition from COBSUB to DEFENDUS criteria in the northern hemisphere. It must be emphasized that the British case is particularly extreme because of the strong commitment to the protection of nuclear power which was the motive that sank Salter's Duck. In contrast, in Norway, where the Duck migrated, and has been taken up, and in Denmark, which has an objective of producing 10 per cent of its electricity from wind power by the year 2000 and which already generates 6 per cent of all wind power in the world, steps have been taken much faster than in Britain.

A striking thing that emerged in the British debate about the future of nuclear electricity in the late 1980s was that once the different fuel sources competed on a level playing field (that is to say once nuclear was compelled to take account of the full costs of its cycle), wind and wave power emerged as not only cheaper in competitive terms but *reliably cheaper* because the variables about their future were known. Objections to onshore wind farms were often proposed by opponents of wind power. In response, Friends of the Earth UK stated that it believed that "wind farms can represent symbols of wise resource use, responsible energy policy and environmental stewardship rather than merely another form of visual pollution."

One of the reasons that Michael Grubb used offshore wind farms as his example was precisely to overcome possible objections to the visual impact.

If, in addition to these inherent advantages on cost, renewable resources were given additional support such as the energy-saving technologies now increasingly being given by American electricity utility companies, their competitive edge in the energy market would increase and the move to address the fourth stage in mankind's energy revolution in the northern hemisphere would advance that much more quickly. What is lacking is only political will.

Developing countries: leapfrogging into the future

But what about the needs of the poor? After all, the Brundtland Commission stated that their needs should be given priority. While super-efficient light bulbs and sophisticated electrical grids could be given to the poor and could be used, these technologies are not those that most directly help them to escape from the poverty trap.

In much of the Third World, people are forced into taking actions to secure their short-term needs which gravely prejudice their own and their children's futures.

In the fundamental area of energy use this is expressed by the fuel-wood crisis. At least 2 billion people – the majority of people in the Third World, rely on fuel wood to cook their food. Yet 70 per cent of them do not have access to secure supplies. For the poorest third of mankind, the collection of fuel wood dominates daily life. As wood is overcut, so increasing amounts of time and labour have to be spent in finding

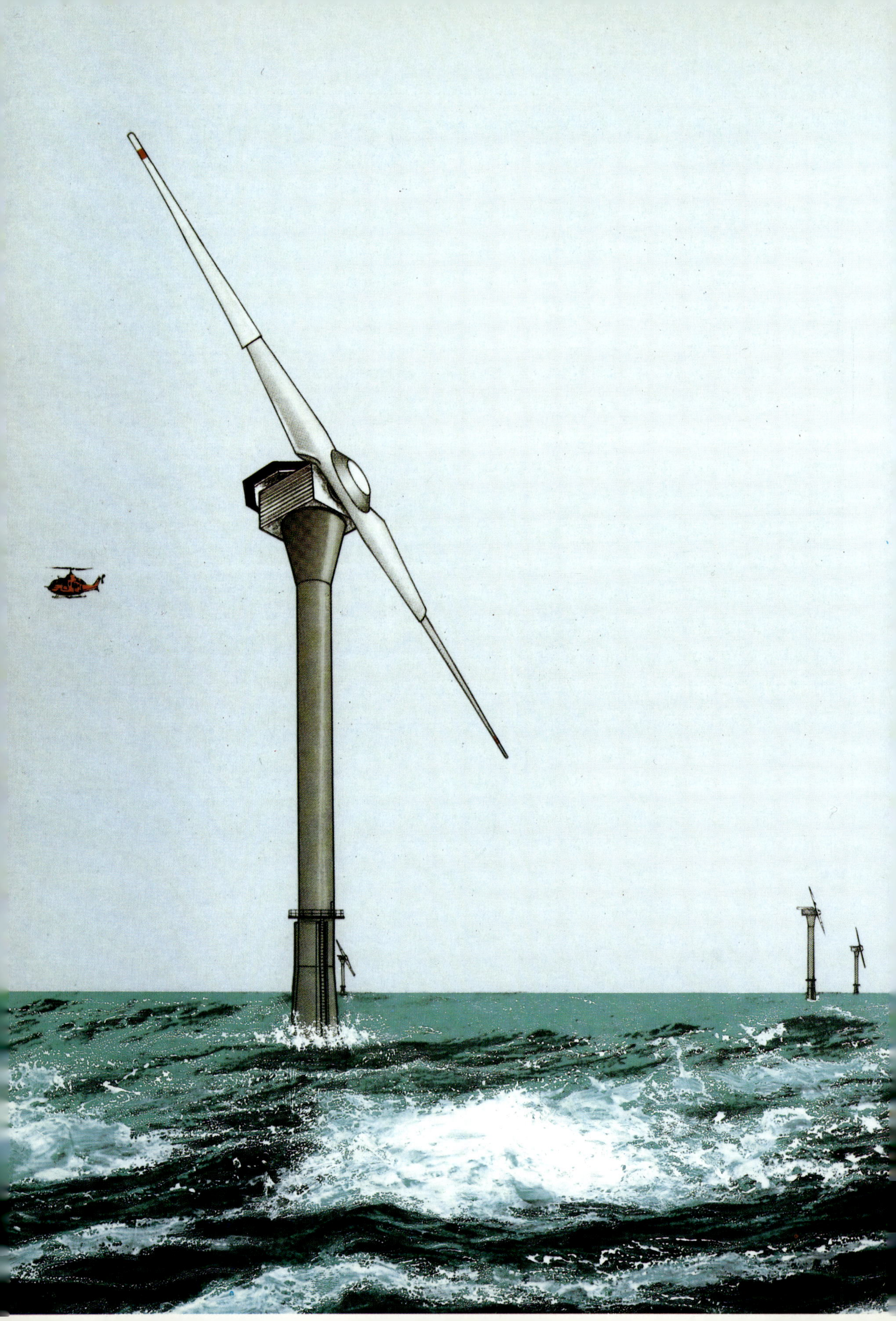

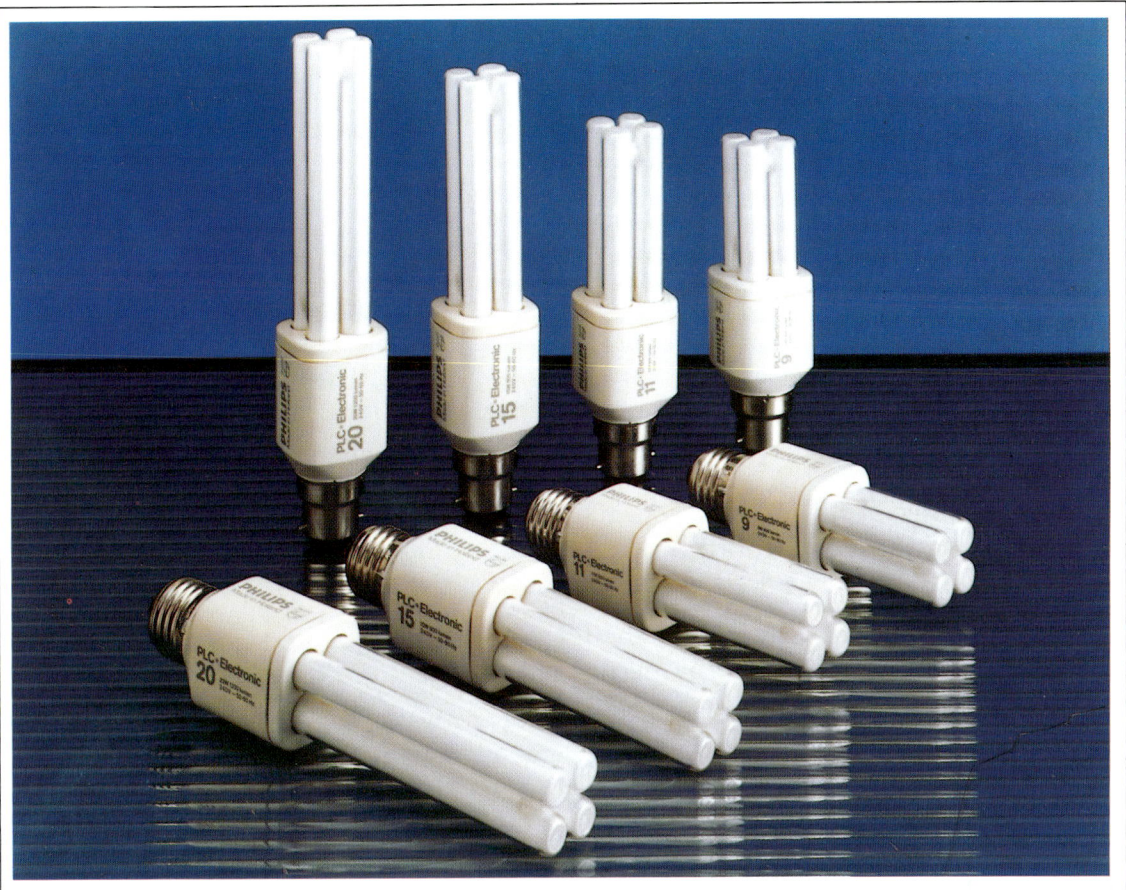

Left: **the potential for off-shore windpower in Europe is vast.** *Above:* **Salter's Duck is an efficient, reliable way to harness the energy of the waves. It was invented in Britain, but Britain will not benefit from it.** *Top:* **The true cutting edge of technology is found not in fusion reactors but in ultra-low energy light fittings.**

> Do we care about the purity of water in an African nation? Do we care about the immunization of a child? Do we care whether a family has a place to live? Do we care if their land is eroded or of it is still able to produce food? Is it in our interest if a stream in Asia still produces clear water and edible fish? Are people living in a state of deprivation, of persecution or unhappiness or uncertainty or are they living a life of relative security?
>
> I think these questions should be addressed in the more affluent nations and the greater and more influential a nation is, the more we can set an example for others to follow. In the long run paying attention to the plight of others assists our own well-being; because when we have a harmonious world we have better trade, better commerce, better profits for our goods as well as better friendships and a more peaceful world. When people are stricken with deprivation, they are much more likely to have conflict. Conflict which can spread and affect our own country.
>
> ... The thing that troubles me most, and is increasingly going to become a matter of unavoidable importance is the chasm that exists between the rich nations and the poor countries. We have plagues like Aids, the plight of refugees and the homeless, civil wars evolving into regional disputes that can then involve the western or northern nations. A refusal by the wasteful countries, which are the rich nations, to conserve oil, for instance, causes a preoccupation with oil and becomes a factor that shapes foreign policy. ... Lack of sharing between the rich and poor nations is a major consideration ... (President Jimmy Carter, December 1990)

it and carrying it home. Half of all the wood cut throughout the world each year is used as fuel wood and four-fifths of that is in the Third World. The fuelwood search has many bitter twists. As more and more people – today perhaps 1.5 billion of the 2 billion relying on fuel wood – encounter problems in finding wood, and as areas become over cut the consequences spread out in concentric manner. The trees destroyed do not regenerate. The removal of tree cover opens up often fragile tropical soils to rain and wind erosion. People try to find substitutes. In Africa and Asia at least 400 million tonnes of dung are burnt as fuel each year. But if applied as fertilizer this dung could help produce many millions of tonnes of food.

The problem is worst for semi-arid tropical and subtropical regions, places like the African sub-Saharan savannah or the Horn of Africa, and it is just such areas which are already stressed which may well be the first to experience the severe consequences of potential climatic change. Professor Martin Parry of the University of Birmingham, who was lead author of the IPCC assessment of potential impacts on agriculture, suggests that places such as western Arabia, the Maghreb, western West Africa, the Horn of Africa and Southern Africa and eastern Brazil, as well as some humid tropical and equatorial regions, possibly Southeast Asia and Central America, could be those most at risk if the IPCC "business as usual" scenario is correct and greenhouse gas emissions continue at present levels. How best then to take action to assist the poor world to climb out of this pit?

From the minimal to the most imaginative, there is much that can be done. The first step is to provide people with better ways or alternative ways of cooking. Better ways can mean the provision of simple, high-efficiency wood-burning stoves to maximize the value of wood burned (40 per cent efficient: double that of a protected open fire). Alternative ways can cross a whole spectrum. One step would be for the wealthy to subsidize the poor by giving them free paraffin or gas and stoves in which to burn it and upon which to cook in return for an undertaking not to cut fuel wood. Put another way, the rich could buy the preservation of tree cover on fragile tropical soils.

This sort of action would be the equivalent of emergency food aid in a famine and it is open to the same criticism: that, in the terms of the Chinese proverb, it is giving a man a fish, not teaching him to fish.

A third step which would move further across the spectrum towards the application of high technology to the tropics would be to take the best of advanced technology and make it modular and usable in the village context.

Since the end of the Second World War, the most pervasive local power source to spread through the poor world has been the diesel generator. But the problem with diesel generators is that they need constant and reasonably skilled care and a supply of pure fuel. It was recognition of the fact that such technology dragged behind it long chains of necessary logistic support, like Marley's Ghost, that helped stimulate the Intermediate Technology movement.

Intermediate or appropriate technology means seeking those types of elegant and relatively simple and robust technologies which are particularly applicable to arduous and remote usage. In the energy field one such example was development work in the 1970s of the external combustion Sterling engine. The engine worked by differential heating in its cylinders and the heat was applied externally rather than injected into the engine and exploded as in an internal combustion engine. Turning slowly, the shaft of the Sterling engine could, for example, drive a water pump. But by far the most exciting possibility to help the poor break out of the fuel-wood trap is at the most advanced end of the technology spectrum.

A photovoltaic (PV) cell is a device which converts sunlight directly into electrical energy by using semi-conductors. The invention of PV systems was greeted with what turned out to be exaggerated optimism. Early PVs in the 1970s had efficiencies of only 8 per cent. Today single crystal and polycrystalline silicon models, which account for about two-thirds of global PV production, have efficiencies of 12–15 per cent and a lifetime of more than twenty years. These technologies are nowhere near as competitive in the industrial world's energy markets as other renewables like wind and wave power.

The best commercial PV systems today produce electricity at about 35 US cents per kilowatt hour. To compete in the American market, this figure must be reduced to 12–15 cents to compete with other energy sources for peaking power and 6–10 cents per kilowatt hour for base-load electricity. (Industry projections are of 12–18 per cent efficient PVs at 20 cents per peak kilowatt by the year 2000.) But for the developing world such competitive costs are not so important. PV systems are potentially a solution to the fuel-wood crisis because they can provide village-size power sources. The electricity generated can pump water, provide lighting, run refrigerators and give power for cooking and do so in an entirely free-standing way not dependent upon any long chains of communication or expensive and unreliable imports.

However, reviewing the state of PV technology in the 1990–91 *World Resources Annual* of the World Resource Institute of Washington DC, the science writer Tom Burroughs suggests that it would be wrong to assess the potentials of PV systems for the sunlit tropical world on the performance of the present generation silicon crystal models.

Two new types of PV technology are now rapidly advancing.

One uses "concentrator" modules whereby lenses or mirrors focus sunlight on small but highly efficient cells. The units track the sun's progress across the sky. Such systems are far more efficient than flat plate collectors and in 1989 a concentrator achieved a record efficiency of 38 per cent. However, in contrast to flat plate collectors, concentrator units are much more expensive to build and do not work as well when sunlight is diffuse. In recent years, however, attention has shifted to "thin film" PV systems in which a semi conductor material is deposited in a layer many times finer than a human hair on an inexpensive material such as glass or stainless steel. These modules are less efficient than conventional silicon models but hold great potential for low-cost mass production. The trade-off is between the low cost of producing the cell and its low efficiency. Burroughs suggests that the future is likely to belong to cheaper low-efficiency systems in large number. He quotes the Director of the US Solar Energy Research Institute who maintains that photovoltaics have the potential to become the primary means of generating electricity world wide by the end of the twenty-first century.

What makes PV solutions to the fuelwood crisis different to intermediate technology ones is that they permit the poor to leapfrog across a whole stage in energy sources. One possibility that PV offers is to match PV cells to electrolysis cells and use the electricity to split water and produce hydrogen for use as an automotive fuel for the twenty-first century. Work on the feasibility of this by Dr Bob Williams of the Center for Energy and Environmental Studies at Princeton University suggests that costs could be competitive with oil-from-coal synthetic-fuel technology and, being a wholly clean fuel, PV systems could have tremendous environmental advantages.

But we do not have to wait for the twenty-first century to see how, in attacking the question of transition from fossil fuels to renewable energy sources, the poor world can leapfrog over the presently industrialized world.

The oil-price shocks of the 1970s knocked many developing countries sideways. This was because they had invested heavily in the decades since the Second World War in automotive transport systems. One of us (Prins) was living in Central Africa at that time and remembers well how cruelly the blow fell upon Zambia, a country which had invested heavily its copper wealth of the 1960s in an extremely good tarred road system and had abandoned old-fashioned wood-burning steam-driven railway power. Yet for that country, as indeed for much of the poor world, railways were and remain the means of mass transport of preference. Prins remembers seeing the then depot manager of the Zambezi sawmills railway standing with his arms outspread in front of one of his ancient steam locomotives. "Look at this!" he exclaimed. "This is the solution to Africa's transport problems! It's a hundred years old, it weighs 97 tons, it runs on wood and water with which the good Lord has filled my country! All it requires is a bit of oil on the bearings. We have welded jacks on to the buffer beams so that you can put it back on the track if it falls off. There's not much on it that cannot be fixed with a two-pound hammer, and it doesn't matter too much if your hit the wrong thing!"

But while many steam buffs around the world, particularly those who flock to South Africa, India and China to see the last great steam-driven railway networks in operation, might sympathize with this Zambian's view, the truth is that a return to nineteenth-century technology on any scale is unlikely.

In Brazil, the oil crisis of the 1970s provided the best example we have yet of technological leapfrogging. With a growing deficit in its trade balance caused by the enormous leap in petroleum prices, Brazil decided to substitute pure ethanol and gasohol (a mixture of ethanol and petrol) in motor-car engines.

The production of ethanol from

fermenting sugar-cane juice rose from 900 million litres in 1973 to 4.08 billion litres in 1981, of which 1.88 billion litres were turned into hydrated ethanol. The remaining 2.2 billion litres became anhydrous ethanol mixed with 20 per cent petrol. In 1989 12 billion litres of ethanol replaced almost two thousand barrels of gasoline a day in five million Brazilian cars.

The alcohol industry created 700,000 jobs. The excellent performance of ethanol-fuelled cars significantly improved the quality of the air in polluted cities such as São Paulo and Rio de Janeiro.

So at a stroke, Brazil was able to leapfrog over the most advanced industrial countries. The fuel-alcohol programme enabled Brazil to run its motor fleet on a fuel which adds zero to the carbon cycle and which reduces many of the unpleasant side effects of motor vehicles upon the quality of urban air. It also meant that Brazil, a developing country, could establish an entire fuel cycle controlling everything from the original source of energy to the devices which finally used it.

The only hitch was that the average cost of ethanol production in the southern region of Brazil was about 18.5 cents per litre. At this price, ethanol could compete successfully with imported oil at an international oil price of about $24 a barrel. So when the oil price fell in the mid 1980s the ethanol programme faced a crisis and only survived as a result of government subsidy. But this Brazil could ill afford and in the face of pressure both to produce ethanol more cheaply and to remove the need for subsidies, ethanol costs have been reduced by 4 per cent per year, and productivity both in the sugar-cane plantation and in distilleries substantially increased.

There is a nice irony about the best of times. When the prescient *Limits to Growth* study was published twenty years ago, many of those who scorned it offered as one of their reasons their belief that the application of advanced technology – nuclear fission and later fusion reactors, solar-power collector satellites beaming high-energy beams back to Earth and so on – would mean that the problems which the doomsters perceived just wouldn't happen. Amory Lovins and other pioneers in the fight for environmental security called these people "cornucopians". They pointed out that these powerful technologies had possibly unknown side effects (many of which have now become known) and unknown costs. For this, they were derided as "technological pessimists".

In this chapter, Professor Lovins and colleagues have appeared in quite another guise. We have argued for the immense, indeed stunning, potentials of highly advanced technology. But not the technologies admired by the cornucopians. Ultra-low-power fluorescent light fittings instead of nuclear fusion power plants; windmills and wave machines under advanced switching systems, not solar-power satellites. The "technological pessimists" are very optimistic indeed, and make the case for being paid to eat the famous free lunch which free market economists are always keen to deny people. Meanwhile, the "cornucopians" have experienced role reversal as well. Lovins's lights, Salter's Duck and the rest don't exist, won't work, aren't cost-effective. And, as we saw, in at least one documented case, if the figures don't support that view, the cornucopians involved manipulated them until they did. People, they say, will be disturbed by the noise of the windmill blades. The cornucopians are the "technological pessimists" now. The role reversal is complete. But there is a difference between the positions in 1970 and 1990. The technologies promoted here are based upon the DEFENDUS, not the COBSUB criteria.

Chapter 6: So What's Holding Us Back?

Let's assume for a moment that we can break out of the sense of powerlessness, that we can overcome the paralysis of denial in the industrial countries which we described in Chapter 3. Let's assume that we can not only make conventional threat analysis work for environmental security threats as we have done in the more traditional realm of threats and enemies but that we can get ourselves to act upon that information. In other words, let's assume that we can create in the industrial world that political will which seems to be a function of increasing wealth.

Crucially, let's assume that the willingness to take action expressed by President Carter can be translated into positive action.

In the developing world, let's assume that the motivation is there to take action to change the "social indicators" of poverty as recognized by the World Bank: to decrease the death rate of children under five; to increase the enrolment of primary school children in primary education; to reduce the number of children per family by raising the confidence of parents that the children to whom they give birth will survive to a fulfilling adulthood; to improve adult literacy and adult life expectancy in the developing world. To the World Bank's pretty conservative list let's add the assumption that there is a belief that gross inequalities of income, such as are particularly found in Latin America, may be reduced. As a consequence of all these conventional development objectives, preconditions can be set to enable people in the poor world to escape from the vicious spiral of poverty and associated environmental destruction by making them able to obtain and use the benefits of intermediate and high technology in association with a rediscovering of the skills and achievements of sustainable subsistence agriculture in the tropics, scorned for three generations.

Let's assume for a moment that the will to achieve all these objectives in the rich world and in the poor world exists. What then holds us back?

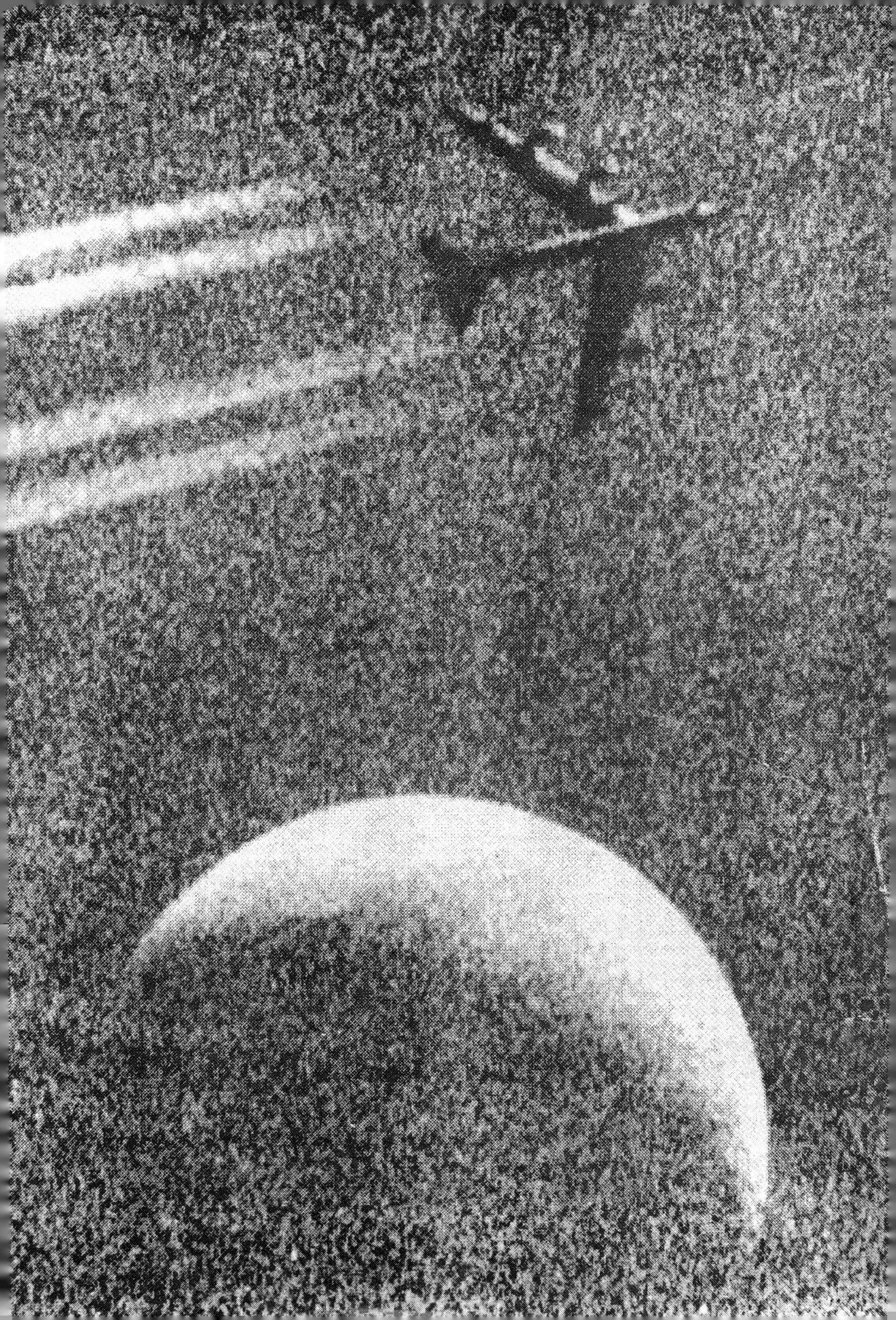

Bolivian families suffer the economic consequences of their country's debt burden.

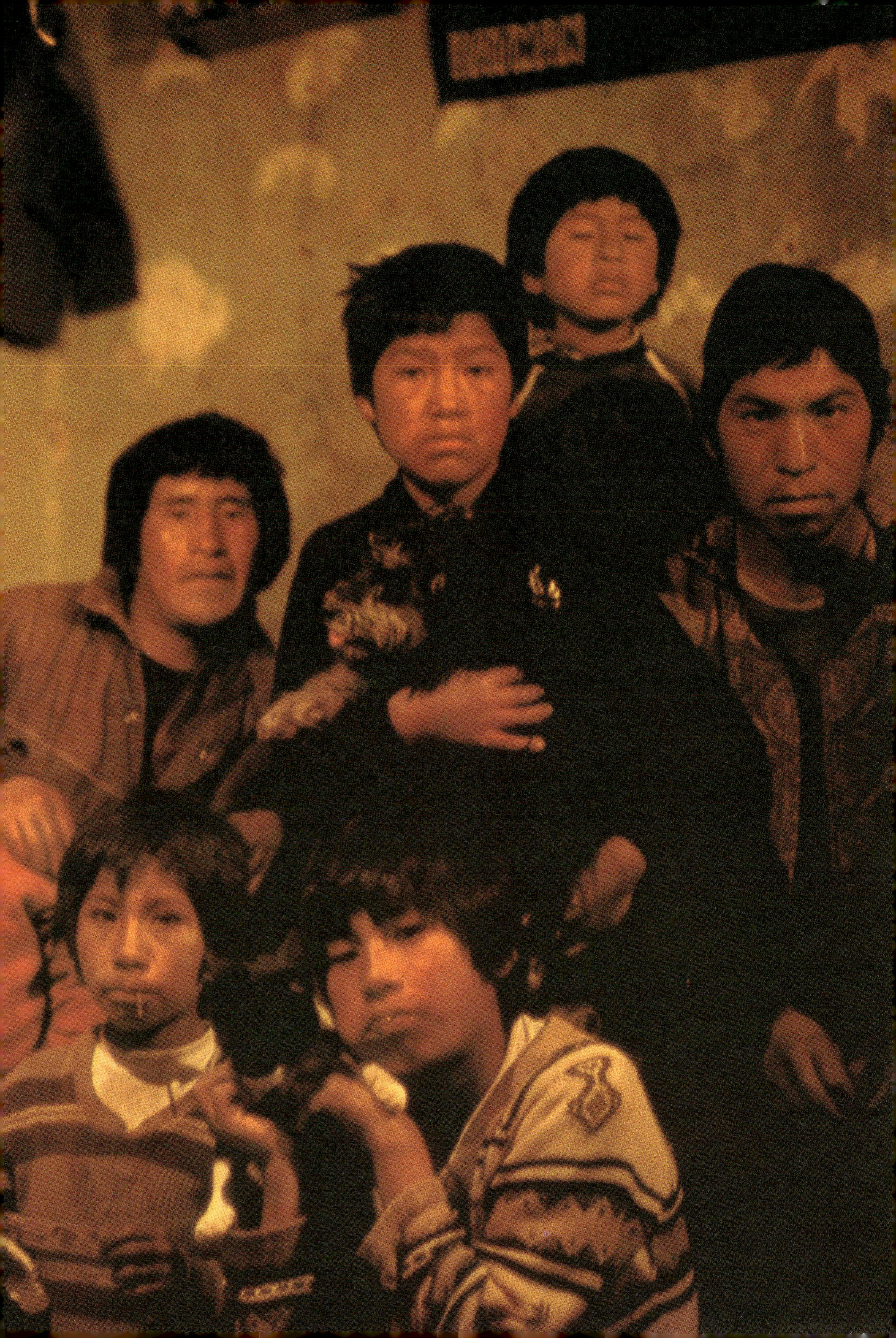

Poor people's chains

Nobody disagrees that the poor and developing world was savagely hit by the rise in energy prices in the early 1970s and that many parts of the poor world, especially the poorest countries which depend overwhelmingly on the export of primary products for their earning power, have been unable to recover from that blow.

But on what should be done about this situation there is wide difference of opinion.

On the one hand, the World Bank takes a rather optimistic view. It accepts that the shocks of the 1970s exposed what it called "structural weaknesses" within developing and poor countries. These, the Bank said, had to be addressed by "adjustment policies". During the 1980s the International Monetary Fund required so-called "adjustment policies" as a condition of lending money to the poor. These adjustments meant the removal of subsidies and protection from prices, which often resulted in sharp rises in the prices of basic necessities, which in turn led in many parts of the developing world to social and political unrest as well as to a great increase of misery. Nevertheless, by 1990 the World Bank was of the view that the medicine had by and large worked. Adjustment, it said, "calls for policies that harness market incentives, social and political institutions, infrastructure and technology . . .", the task of making the labour power of the poor more productive. The second strand to "adjustment policies", the Bank thought, was "to provide basic social services to the poor. Primary health care, family planning, nutrition and primary education . . .".

In the second area, the Bank believes that those who saw the 1980s as a "lost decade" for the poor are not right. Because, with the major exception of sub-Saharan Africa, the Bank perceives a trend of general improvement in all the major social indicators. What is needed now, the Bank believes, is to continue this process of helping poor countries to become members of good standing in the international market economy and that means in particular that those countries carrying burdens of international debt must find ways to pay off their debts. The discipline of doing this will do them good. To this end, the World Bank instituted a plan (the Brady Plan) which had as its objective lending money to indebted countries to help them undertake actions to become less indebted by being full members of the international economy.

But many critics of the World Bank approach do not believe that this strategy will work either in its own terms or will help address the new agenda within the points of the triangle of development, environment and resources.

> The most important condition for sustainable development is that environment and economics be merged in decision making. Our economic and ecological systems have become totally interlocked in the real world but they remain almost totally divorced in our institutions. (Jim MacNeill, Secretary General of the World Commission on Environment and Development.)

The critics simply don't recognize the picture of the poor world painted by the World Bank. Geoffrey Lean, Don Hinrichsen and Adam Markham write quite bluntly in the World Wide Fund for Nature *Atlas of the Environment* that "for hundreds of millions of people, the 1980s were a lost decade. Development stagnated, and often regressed, in much of the Third World. A third of the entire population of the world live in countries which experienced either zero growth or actual decline during the decade."

Why was that? Was it simply that the

poor were so much less skilful than the rich at managing the shocks to the international economy? Certainly the World Bank sees the root of the problem in the Third World: "... increased capital flows can provide only temporary relief. They are no substitute for domestic action ...". But that is not the critics' view.

The sudden rise in oil prices at the beginning of the 1970s meant that the oil-rich nations of the Middle East became the recipients of the largest single outflow of money from the industrialized world to any one region to have occurred since the First World War. What on earth were they going to do with all these petro-dollars? If they simply built up as savings there was a real threat that the entire international banking structure would collapse, capsizing like an unstable ship that is holed in an accident.

The commercial banking structure came up with a brilliant solution. It offered the oil rich the opportunity not only to deposit their money but to make their already large earnings grow more by interest because the banks would re-lend the oil money to those who needed it; and since the lending would be in the form of "sovereign debt" there would be no risk, because, bankers believed with that trusting naivety which, John Maynard Keynes observed, is one of the marks of their profession, that while an individual can go bankrupt because of profligate practices or unwise investment, by definition *a country cannot go bankrupt*. This splendid view, dignified with the title the "sovereign risk hypothesis" was associated with Mr Walter Wriston of Citibank, which in the later 1970s set the pace. Bankers had another reassuring device, used also by sheep. It was the "syndicated loan" whereby each bank tied its participation to that of each other party. This made them feel safer.

So huge amounts of money as syndicated sovereign loans were pumped out into Latin America and Eastern Europe, and to a lesser degree (but very important for the borrowing countries) to the poorest countries, to help them pay for the sudden increase in the cost of their imports of energy and other goods without any consideration at all being given to whether the loans were properly secured against a collateral – it wasn't necessary. The banks vied with each other to see how much debt they could shift each month, and published lists, charmingly called "tombstones", to show who had done best. When countries started to default on their repayments, the loans were "rolled over". Rolling loans gather no loss, but mount up alarmingly.

Having borrowed the money, of course the poor had to repay the principal and interest. Their source of earnings was, as we have just pointed out, very largely raw materials, and in many countries an overwhelming dependence upon just one or two crops or minerals. Yet, as a result of the world recession, demand slackened and the terms of trade in the 1970s and 1980s moved sharply against these exporters. So their sugar or sisal or copper was worth less and less while their debt burdens increased more and more.

The World Bank and IMF policy was to try and encourage those activities in the developing world which would maximize their foreign currency earning potential so that these debts could be serviced. The result has been a grotesque reversal in the flow of resources from rich to poor. At the beginning of the 1980s, the poor received at the end of the day a positive flow of about $18 billion once debt repayments had been set against the inflow of aid in various forms from the rich world to the poor.

Between 1982 and 1988 developing countries repaid $830 billion in interest and principal, which was far more than the debt they originally owed, and still their debt burdens have continued to grow.

In 1983 the inward flow of resources to the poor world became a trickle and in 1984 it reversed. From then until now, in a grotesque inversion of what most people in the rich world might expect, the financial flows have rapidly accelerated *but in the opposite direction*. The poor transferred $10 billion to the rich in 1984. This had become $44 billion by 1989.

With the greatest relative burdens of debt being borne by the poorest countries, and with ever increasing amounts of their foreign currency earnings being spent upon debt servicing, it was hardly surprising that progress on the "conventional development social indicators" (such as those recognized by the World Bank) was slow and action on those areas where additional expenditures are required to achieve sustainable development was minimal. A quarter of sub-Saharan Africa's earnings from exports between 1985 and 1987 went to service debts, but even so met only about 60 per cent of obligations.

In contrast to the World Bank, the World Resources Institute in 1988 calculated the additional expenditures which would be needed to make progress in target areas, prominent in which was the development of "leapfrogging" energy technologies for the Third World. Like an increasing number of bodies and commentators, the World Resources Institute takes the view that the time has come to release the poor countries from the shackles of inherited debt incurred initially during the early 1970s.

Among those also taking the view that these debts will never be repaid are many of the major banks who were the primary lenders in the first place. In the late 1980s, the nine major "money center" banks of the United States (names like Chase Manhattan, Chemical Bank, Citibank, Manufacturers Hanover) and the big lenders in the UK like the Midland Bank and National Westminster set aside huge amounts of money against "bad debt" and reporting losses, often for the first time in their trading histories. In the mid 1980s, Midland had the dubious honour of being the most exposed of all the Anglo-American banks (205 per cent of capital on Latin American debt alone in 1984). One, Continental Illinois, actually went bust, but was promptly nationalized by the Reagan administration (ideological objections deftly set aside) which greatly reassured the other over-exposed banks.

What many development and environment experts now propose is a conscious policy of graduated default on these debts. They do not share the World Bank's view that there is some value in making poor countries learn from the "discipline" of repaying debts incurred in the way in which these were. We might ask whether they are correctly described as debts. In one sense they clearly are not.

The money that was "lent" to the Third World in the 1970s was less a loan, more a price that was paid for a product. That product was the *non-collapse* of the international financial system. So, using free market logic, we might reasonably argue that since every commodity has a price, the price of this non-collapse is the cost of relieving Third World debt. The alternative might have been for the industrial world to plunge into deep depression. Thus we could say, with justice, that relieving this debt is cheap at the price.

Some campaigning organizations like Friends of the Earth have been arguing that the close link between environmental security and the economic activity of the developing world could be beneficially strengthened in the process of relieving this debt by making "debt for forest" swaps. Indebted countries containing critical assets like tropical rainforests might effectively be paid to cease the destruction of those forests because logging of tropical hardwoods as a means of earning foreign currency to pay debt servicing payments will no longer be necessary.

"But", it will be immediately asked in the rich world, "where is the money to come from to pay for this relieving of the debts of the poor or indeed for all the other environmentally friendly things that need to be done?" The answer to the first is easy: it has already been found by banks writing off the bad debts in their books over time. The answer to the second we now address.

The chains of the rich

The political obsession with military force which has dominated the priorities of countries since 1945 has sapped not only wealth but also the world's pool of skill and inventiveness to a dramatic degree. From 1960 to 1987 Ruth Ledger Sivard in *World Military and Social Expenditures 1989* calculates that we have spent $17 trillion on a global arms race: 70 per cent more than the $10 trillion which the world spent on its health budget during that period; 13 per cent more than the $15 trillion spent on education. So, in one sense, just as the chains of debt have hung heavily on the poor, so the chains of military expenditure have hung upon the rich.

The two superpowers of the post-war confrontation, comprising about 20 per cent of the world's population, have had in their alliances until the end of the Cold War about 40 per cent of the world's armed regular soldiers and accounted for over 80 per cent of all military expenditures.

Spending reached a peak in real terms in the early 1980s. In the American case, the last four years of the 1980s saw a clear downward trend from the high point of the Reagan years, while data available to the Stockholm International Peace Research Institute for Soviet military expenditure in 1989 shows that for the first time in two decades defence spending slowed down over the preceding three years. In 1990 Georgy Arbatov, Director of ISKAN (the Institute for USA and Canada Studies) and a senior Gorbachev foreign policy adviser, asserts that it was reduced in real terms from 77 to 71 billion roubles. Against this, however, must be set the fact that, with Mr Gorbachev's move to (or capture by) conservative forces in the USSR in January 1990 and his willingness to use armed force against the Soviet republics, Soviet military spending has been budgeted to rise by 30 per cent in 1991 to 96.5 billion roubles, representing 37.5 per cent of all Soviet government expenditure. By way of comparison, the education budget was set at 4.6 billion roubles and health at 2.9 billion roubles. All this presumes that the Russian Federation will actually reverse its decision to cut massively its contributions to central government.

However, overall, what we see is a picture for the industrial world of an ending of the rapid increases of the early 1980s but still no sign of deep cuts in defence spending. Countries seem to be following a policy of "wait and see". As we go to press, we do not know what will happen as a result of the military confrontation in the Gulf or from the use of force against the dissident Soviet republics in terms of the trend in stabilizing and possibly reducing world military expenditures.

We do know already that the crisis since August 1990 has had the effect of producing enormous contracts for arms importation into the Middle Eastern region. Sales to Saudi Arabia and Israel may signal a change in a trend to be seen over the 1980s in the international arms trade.

Here we had seen the high growth of exports in arms to the Third World in the 1970s replaced by a trend of shrinkage in the value of the global trade in conventional arms since 1982 fluctuating between $30 billion in 1980 to $39 billion in 1987, a peak year. Within this trade, the share of the industrialized countries steadily increased against that of exports to the Third World: 33 per cent in 1987, 42 per cent in 1988 and 50 per cent in 1989. In contrast, imports of major conventional weapons to the Third World in US dollar terms fell to $16 million in 1989 – the lowest level since 1976.

The accusation often made against countries of the Third World that they would be better able to help the poor if they spent less on arms, while true in absolute terms, is less true in relative terms than critics might think. Those in the wealthy industrial world should first remove the beam from their own eyes before trying to remove the speck from the eyes of the other! That said, the speck is pretty horrible, for the trend has been increasingly that most of the military equipment bought by poor countries outside the Middle East is for use by governments against their own people.

The case against the international arms trade is not just the absolute diversion of resources represented by military spending world wide. It is that these expenditures appear to work in a general and systematic way against the fastest realignment of advanced technology towards the objectives of sustainable development and environmental security.

The United States ranks first in the world for military expenditures, first for military technologies, first for the number of worldwide military bases, first for military training programmes for foreign forces, first in military aid to foreign countries, first in the size of its naval fleet, first in its numbers of combat aircraft, first in its number of nuclear reactors, first in the number of nuclear warheads and bombs, first in the number of nuclear tests, second in the scale of global arms exports (although first in terms of arms exports to the industrial world) and third in the size of its armed forces among the 142 countries in the world. In contrast, while being first in the percentage of its population with access to safe water and the number of births attended by trained personnel and second in gross national product per capita, the United States is fourth in female and male literacy rates, sixth in the number of school-age children in school, thirteenth in maternal mortality rates, fourteenth in life expectancy, fifteenth in the percentage of population with access to sanitation, seventeenth in its infant mortality rate, twenty-first in its under-five mortality rate, twenty-fourth in the proportion of population to physician, thirty-first in the percentage of infants with low birth weight and thirty-second in the percentage of infants immunized against measles.

In making these dramatic comparisons, Ruth Ledger Sivard intends to make a simple but important point. That there is here an *opportunity/cost* choice being made. Being first in terms of military power seems to impose a cost in terms of social development. The same case can be made even more strongly for the Soviet Union which, while almost matching the United States in military expenditure, did so at much greater proportional cost to its total economy.

But again, the choices made to pursue the old instruments of international prestige and power are not only choices against advance in conventional social development. For, as the American economist Jeffrey Dumas has explained in his book *The Overburdened Economy*, not all expenditures are the same. Some he calls *contributive*. These are investments which produce greater social or economic wellbeing. Some are *neutral*. They do nothing to contribute anything to a final product and consist of featherbedding, paper pushing, the provision of useless information and so on. And some expenditures (among which he includes most military expenditures) are *distractive*. (For economists, Professor Dumas's taxonomy can be recognized to be a modern restatement of Adam Smith's distinction in *The Wealth of Nations* between productive and unproductive labour.)

How can military expenditures be viewed as distractive and ultimately damaging of long-term economic wellbeing, especially defined in the broader sense of sustainable development? In one obvious sense. Military forces are designed for the express purpose of destroying as efficiently and as widely as possible, productive assets and people. To this, in modern warfare, must be added the recognition that the environmental side effects of combat can far exceed in long-term effect the calculus of profit and loss upon which the decision to go to war was traditionally made. But military expenditures are distractive in two less immediately obvious ways also.

First, there is a limited pool of skill. The more talented engineers and technologists there are who spend their lives designing clever ways to make missiles go in wiggly lines, the fewer there are to devote the same skills to improving the energy efficiency of appliances or designing better forms of power generating windmills.

This is only common sense. Thus, when we look at the economic performance of major industrial countries since the end of the Second World War, we see that the two countries which lost that war and which in consequence had their military budgets strictly curtailed and which devoted their scientific and technological skill predominantly to the improvements of civil goods – that is to say West Germany and Japan – achieved during those years the most spectacular economic growth, while establishing themselves as producers of goods of high quality.

It is therefore ironic in the light of this convincing demonstration of what can happen if the opportunity/cost decision is made *against* military expenditure, that in 1989 Japan was the largest single importer of American arms and, with an expenditure of $3.1 billion, the second largest arms importer in the world, surpassed only by India. It is really a double irony because the wealth which Japan uses to purchase these arms is the product of its remarkable growth in the world economy which has resulted from *not having manufactured arms* in the preceding generation.

It is often said that the argument that military expenditure has a negative effect upon general economic progress is false, because there is a demonstrable "spin off" from military technologies, or from the needs of warfare, which benefits the civil sector. We have already mentioned how the two world wars hugely propelled the advance of the internal combustion engine on land and in the air; and out of the Second World War came the jet engine and atomic power. In an "action/reaction" cycle, there followed the superpower arms race which included the space race, and out of this came rockets whose nose cones required Teflon to slip through the atmosphere. Teflon was subsequently applied to the inside of frying pans. The point is often made that if you want to develop a nonstick frying pan it seems odd to develop a missile nose cone as a means of doing it.

But the fact is that, standing now at the end of the East–West Cold War, it can be seen that huge expenditures have been made. Whether or not these were sensible this is how the world has chosen to spend its time and wealth. The challenge is to find whatever means we can to draw benefit in as many ways as we can from those military expenditures and to apply that benefit to the attack upon the new enemy whose existence has been thrown sharply into focus when viewed in terms of military threat analysis.

Chapter 7: Captain Cook to the Rescue

A transitional world

We do not yet know whether it will be the best of times or the worst of times. We can see that the motivating venom has gone out of the superpower confrontation as the Soviet Union passes through the second Russian Revolution. But we can also see that threats of the old-fashioned sort, defined in "defence" terms, have not gone away. The way in which these "defence"-defined threats show themselves is strikingly different from the frozen deployments of armour and aircraft staring beadily at each other across the ceasefire lines of the Second World War in Europe.

The military role which has rocketed to prominence during 1990 has been that of internationally sanctioned blockade and enforcement under the terms of the United Nations Charter. With the end of the Cold War, the United Nations as an organization has, like the sleeping princess, the potential to come back to life. Those parts of its Charter, the Chapter VII powers, which deal with breaches of the peace and acts of aggression have, as a result of the 1990 Gulf crisis, been brought rapidly to the centre of attention.

The deadline for delivery of this book was 15 January 1991, set by the publishers before the UN set the same deadline to Saddam Hussein to leave Kuwait or face the possibility of war. The book was two days late, going to press as the missiles and bombs struck the Iraqi military-industrial complex. But, as we said when we visited CASSANDRA in Chapter 2, a resource war over oil may not be the worst that we have to fear. We would suggest that in the long term, the unprecedented high-technology, high tempo air campaign in the Gulf and the deployment there of much of the sophisticated military hardware designed for battle with the Soviet Union during the Cold War does not provide a model for the most likely sorts of warfare in the future. However terrible the consequences of a

A dolphin caught in a drift net. Each year millions drown in these nets, which trawl huge areas of ocean. As the practice has been outlawed by many governments, navies might well find themselves called on to enforce the law.

war over Kuwait may be – in loss of life, in polarizing a North–South confrontation, in plunging the world into depression as a result of the oil price increasing greatly, or worst, in ecological disaster should vast amounts of oil be spilled or ignited as a result of hostilities – it is unlikely to be typical of the conflicts which could occur in the worst of times.

So it is more important, even as war storms in the Iraqi desert, to plan for the use of UN powers in peacekeeping and preventative roles, rather than for intervention.

At a lower level of potential violence but of longer term potential significance is the prospective extension of the peace keeping powers of the United Nations forces from roles of separating hostile states into protecting the global heritage.

Here again there are sleeping powers within the United Nations Charter which could be awakened, notably the Chapter XII powers relating to trusteeship over former colonial territories (powers which applied, for example, to Namibia before its independence).

> **Chapter XII powers under the UN Charter**
> There may be designated, in any trusteeship agreement, a strategic area or areas which may include part or all of the trust territory . . . (Article 82)
> All functions of the United Nations relating to strategic areas, including the terms of the trusteeship agreements and of their alteration or amendment, shall be exercised by the Security Council . . . (Article 83.1)

The trusteeship clauses can be applied directly to Antarctica. They require amplification before they can refer to areas already members of the UN. An interesting Soviet proposal has been to use these clauses to define areas critical to the health of the biosphere as being special strategic areas to come under Charter protection in a general stewardship of the United Nations exercised through the Security Council. Were this to be done, then the full apparatus of the international community's protective potential, including its military potential, could be deployed to police and protect such areas.

A particular and important initiative in this regard, which would require development under the Law of the Sea regime, would be to encourage seafaring nations to contribute naval forces to the creation of a United Nations Standing Naval Force (UNSNF) which would in normal times operate as an "ocean guard" that could patrol the high seas under a mandate from the global community, just as coastguards patrol territorial waters, but in contrast acting to deter or stop activities which it is beyond the competence of any single littoral state to act upon.

A good example would be the potential use of United Nations mandated forces to stop the dreadful use of "wall of death" drift nets in the open seas. Australia has used its military forces to prevent the use of such nets within its territorial waters; nets often 30 miles long which drift through the oceans trapping every living thing that swims into it. Then there are incidents of pollution in the high seas and, once the Law of the Sea has been renovated and developed to accommodate it, the tracking and signalling of the sources of shore-based pollution.

Moving on from the familiar world of exclusive military alliances in opposition to one another to the creation of international police forces tasked, in particular, for the protection of critical areas would be a significant symbol. It would also mean the legitimate retasking of military forces, which would thus acquire environmental-security-related roles in addition to residual military-defence roles. In a transitional world, until such time as we can create a working discipline for regional and global security systems appropriate to the new age, it is very likely that there will be many incidents in which armed forces will be employed in a "protective" and traditional sense.

The dissolution of the distinction between combatant and civilian which has been an insidious trend since the American Civil War, and especially since the coming of the nuclear age, means that hostage taking and acts of terrorism are likely to increase. In such circumstances, nations will certainly respond by increasing their provision of highly trained, highly mobile "irregular" and special forces able to act with suddenness and precision, as the Israelis did when they rescued their citizens held hostage at Entebbe Airport in Uganda in a spectacular long-range operation.

Another sort of problem which is likely to increase before it decreases and which is likely to involve palliative military reaction (for there are no military solutions to any of these issues) is that of refugees. Whatever the rights and wrongs of the situation, public opinion in the rich world is likely to demand strengthened border patrols, such as we saw in the Mexican case study in Chapter 4. Already, the "war" against drug-runners in Central America has witnessed a worrying elision of police and military roles, with the armed forces being used in active support of drug enforcement agencies.

Of course there are many in the military world who would be happy to argue for the validity of all these so-called "low-intensity" roles as a way of protecting their jobs and budgets; but that is a weak argument, and one easily overridden if the general trend of public policy is away from bloated military establishments which, the Gulf War despite, it is.

None of these continuing military roles in a transitional world calls for anything like the vast expenditure of resources and human effort which created the great mountains of armaments and vast arrays of armed force of the Cold War confrontation. Naval, amphibious and special forces, with appropriate aircraft – more of the maritime patrol and transport types than the highly advanced high-speed fighter or bomber – are a far cry from the world arsenal described in the 1990 *Military Balance* published by the International Institute for Strategic Studies.

But that is not the situation now. The situation now is that the force structures of the previous age still exist and, both in the public mind and in the policy makers' minds, no convincing reasons have yet appeared to do away with them. It is true that the Conventional Forces in Europe (CFE) Arms Control Treaty offers the prospect of the deepest cuts yet achieved by an arms-control process in the non-nuclear weapons field. But that treaty does not touch naval forces, or indeed the other armed forces of the signatory nations stationed outside the treaty area bounded by the Atlantic and the Urals, and serious question marks hang over Soviet compliance at the end of 1990 which may yet stall the initiative.

As we saw in the last chapter, the global trend in military spending does not yet reflect the view that the ending of the Cold War permits a large drawing down of forces; and the sudden sucking in of external powers to the maelstrom of conflicts in the Middle East may yet serve to halt the trend towards deep cuts for a time.

Yet at the same time that we view this transitional world, with all the doubts and hesitations which mark security policy, the case for action upon the environmental security agenda becomes ever stronger. The analysis in CASSANDRA made that clear and the applicability of the medical model to questions of the planet's health confirmed it. Is there no way at all in which the vast resource, both of skill and machines, represented by the world's military can be applied to the new security agenda at the same time as public opinion and the military themselves continue to argue for the continuance for the time being of military capabilities appropriate to dealing with threats which have identifiable enemies behind them?

Immediate needs; immediate threats

Their Lordships the Lord Commissioners of the Admiralty really knew how to butter up the mad King George III. "Shortly after his accession to the throne, having happily closed the destructive operation of war, he turned his thoughts to enterprises more humane, but not less brilliant, adapted to the season of returning peace," they effused. This was how the account of the famous voyages of Captain Cook was introduced. "While every liberal art and useful study, flourished under his patronage at home," they continued with a description that many contemporaries and most subsequent historians would find hard to credit, "his superintending care [sic] was extended to such branches of knowledge as required distant examination and enquiry; and his ships, after bringing back victory and conquest from every quarter of the known world, were now employed in opening friendly communications with its hitherto unexplored recesses".

Cook's voyages were of crucial importance in expanding Europeans' geographical knowledge into the Pacific. Carried on board HMS *Endeavour* was the gentleman scientist and explorer Joseph Banks on behalf of the newly founded Royal Society. Cook's secret orders were to offer him every assistance in his observations and explorations. The ship was both a fortress and a floating laboratory from which Europeans viewed the new world into which they sailed.

Cook's first voyage was instrumental in increasing knowledge not only of the flora and fauna and anthropology of the Pacific but also in advancing the scientific method of study. That it was deemed a perfectly proper purpose for the Royal Navy of that time in "an era of returning peace", was signalled by the Lord Commissioners' sponsorship of the publication of the account of the voyage.

> There's no doubt in my mind that environmental factors are important, and I would say probably very important, in the overall security context. You have only to look at the effects of erosion, drought or trans-boundary pollution to see obvious sources of conflict, and we've seen many of these in the world already. Environmental factors are important and probably will become increasingly important in a global security context. (Admiral Sir Julian Oswald, First Sea Lord)

In 1990 the scientific working panel of the IPCC stated that, while it believed that the general shape of the science of climatic change was now grasped, there was need for a rapid expansion in the volume of data to refine the Global Circulation Model of the climate. In its first report, the Hadley Centre for Climate Prediction and Research, opened in May 1990 as a centrepiece of British research effort, with a primary mission of building and coordinating a three-dimensional Global Circulation Model by the end of the millennium, detailed its proposed collaborations and invited further partners. IPCC Working Group One had proposed the "business-as-usual" scenario on cautious grounds, and the prediction was very worrying. There is a possibility that the result of global warming could be worse than that predicted because, until later, the Global Circulation Model will not be able to model with accuracy the possible acceleration to warming which could arise from positive feedbacks. One worrying aspect of present knowledge is that more, and more potentially powerful, positive rather than negative feedbacks (which might act as counterbalances to warming) are identified.

During 1989 much store was set by some scientists upon the likely albedo

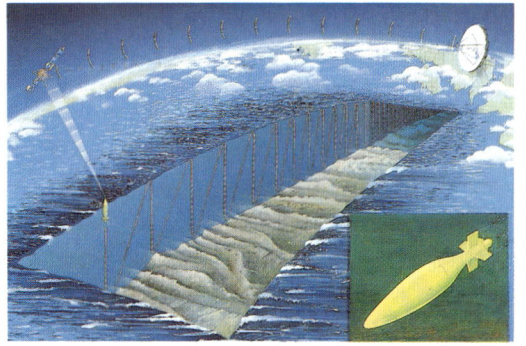

The Rolls Royce gas turbines (*main picture*) which power HMS Cornwall, a Royal Navy Type 2 Destroyer, can also drive small power stations to give carbon-free electricity generation, using bio-gas. The NERC autosub programme (*bottom left*) will greatly increase knowledge of ocean currents and composition. *Bottom right:* navies which run nuclear submarine fleets all possess information which will be invaluable in reducing uncertainty about global warming.

effect which might arise from an increase of cloud cover in a warming world. The white cloud would reflect more solar radiation and thus off-set temperature rise at the surface. This is the most discussed of the identified negative feedbacks. On the other side, the potential for runaway heating should the vast frozen methane deposits in cold latitudes start to be released in gaseous form is not yet in the global models.

During 1990 the range of uncertainty was made plain in a scientific debate about the degree to which we can rely upon the cold oceans to act as "sinks" for carbon dioxide, absorbing CO_2 from the air and locking it up in denser form in the oceans. The suggestion was made that we could, perhaps, rely upon the cold oceans to lock up less CO_2 than had been previously assumed. Should this turn out to be correct, it would mean that the earth already relies more than we think on the land buffers – principally the tropical rainforest – to capture CO_2. It is rather important to know the answers to these sorts of security questions quickly.

Crucial to survey widely are the cold oceans of the high latitudes; and yet these are areas in which it is expensive and difficult to place survey ships and, in any event, to obtain large numbers of accurate readings of a wide range of observables. However, these are areas in which the warships of the Royal Navy and other navies, built for other purposes during the previous generation, still operate extensively.

Would it not be sensible, then, to add the making of the relevant scientific observations to the range of duties of such ships as they went about their other business? Provided that the scientists were able to obtain such readings at the marginal cost of making them – in other words not charged the full cost of hiring the vessel – such a modern version of Captain Cook's voyages would offer a concrete example of how, during the transitional phase, existing military assets designed for "defence" roles could simultaneously be deployed in the front line of attack upon the environmental security problem.

The newly established United States Global Change Research Program run by Dr Eric Hartwig, Director of Ocean Sciences in the Office of Naval Research, has worked out a set of criteria to make such dual-purpose tasking feasible. Both traditional defence-planning criteria and Global Change Research Program criteria were evaluated in parallel as each mission was examined. This was done in coordination with the Committee on Earth and Environmental Sciences (a Federal Interagency group), with the National Academy of Sciences Committee on Global Change and with the various branches of the armed forces. From this process, "candidate areas" were identified: High Latitude science and engineering; data management; observations of the global carbon cycle; modelling and forecasting. Specific projects were adopted by the different services: the Heard Island acoustic tomography experiment by the Navy*; permafrost dynamics by the Army; ice-lead dynamics and ice-thickness studies by the Navy.

The Royal Navy has also begun to investigate how it can add environmental-security research to its missions. In July 1990, the Global Security Programme in the University of Cambridge, assisted by the

---

*What this means is that, starting in January 1991, two American ships, the *Amy Chouest* and the *Cory Chouest*, positioned in the Southern Ocean off the Antarctic coast will generate a loud, low-frequency hum at precisely fixed moments in 300 metres of water near Heard Island. Heard Island is chosen because all the world's oceans can be reached by the sound. Receiving stations on the Atlantic and Pacific coasts of the USA and on every continent will measure the time that it takes for the sound to reach them. Since sound travels faster in warm water, it is hoped that the experiment will enable fluctuations in ocean temperatures to be measured. It is possible that the hum, which is close to the frequency used by whales for long-distance communication, may harm them. But on the other side of the case, global warming, which the Heard Island experiment measures, may harm them more in the long run.

American Academy of Arts and Sciences in Cambridge, Massachusetts ran an initiative with the help of several environmental scientists, research organizations and with the participation of the British Ministry of Defence and the American Department of Defense to try and see whether one could go further than the USGCRP in defining exactly what environmental-security research opportunities for navies might look like in the maritime environment.

Several kinds of significant contribution, which could be made at little or no additional cost to the Royal Navy (or in the course of time to other participating navies), were seen.

There are four stages of increasing involvement (and cost).

Turning to Face the New Enemy

■ Stage 1: Making available data from existing sensors on existing ships and aircraft following existing operational patterns.

■ Stage 2: Use of existing assets following existing operational patterns to carry additional automatic recording or sampling equipment requiring little or no attention by service personnel.

■ Stage 3: As 2 but using environmental observational requirements as a consideration in weighing the balance between training options which are from an operational point of view equally attractive.

■ Stage 4: Use of assets including shore assets to support environmental-research activities in ways which would involve additional expenditure.

Here are some examples of what could be done.

■ The Scott Polar Research Institute in the University of Cambridge has already undertaken vital work on measuring the thickness of Arctic ice from nuclear submarines of the Royal Navy. Nuclear submarines are the only vehicles able to get under the ice to make these sorts of observations, and these sorts of observations are vital to discover what changes are occurring in the extent and nature of the Arctic icecap. Ice data are critical in the study of global warming. In September 1989, the Scott Polar issued a report based on sonar measurements made by British nuclear submarines which showed that in an area north of Greenland the icecap had thinned from an average of 6.7 metres in 1976 to 4.5 metres in 1987. This sort of evidence may still be scientifically anecdotal. It is important to know.

Furthermore, large amounts of information about the conditions of the ocean and of oceanic weather are collected by navies as a routine part of studying their operational environment. Much of this information is already used, for example, by meteorological services in the making of weather forecasts. But the possibilities of releasing more data and different sorts, for example, data on chlorophyll fluorescence as a means of studying microorganisms in the sea, could be undertaken.

■ At Stage 2 new sensors could be mounted to obtain rapidly many observations of crucial variables. Scientists believe that one of the most important of these is the measurement of the partial pressure of CO_2 in water which can be done by automated underway sampling. It is expected that equipment to make such samples will be ready within one or two years and will be suitable for fitting to all types of surface vessels.

Also by automatic underway sampling, nitrates, phosphates, silicates and oxygen in the layers of the ocean can be sampled (by submarines) and with new equipment, submarines can also measure minute amounts of artificial substances deposited in the oceans which can be used as tracers in the mapping of ocean currents. Two substances in particular which should not have been released into the environment but have been can be used in this manner. They are CFCs and the radioactive isotope tritium, deposited in the biosphere as a result of atmospheric nuclear tests. The life of the oceans can be studied through the analysis of plankton. Two techniques are important here. High-frequency acoustic detection of plankton in the sea which permits the mapping of the plankton, and the study of plankton by the use of gene markers.

■ In addition to these opportunities for on-board measurement, the military possess large and potentially very valuable banks of data which the scientific community could use to advance the study of environmental security.

The first type, and potentially the most important type for the scientific community, would be the release of time series data on changes in the Arctic ice. The problem here is to ensure that the operational purposes for which these data on the ice were first collected are not, in the view of the navies who possess the data at present, jeopardized. Thus a process of "sanitization", which means ensuring that, for example, the tracks which submarines follow under the ice are not revealed, must take place.

A second type of data which can be immensely valuable can be obtained by giving access to informal bird observation records made by Royal Navy officers over time which permits the studying of birds' responses to the changing environment.

A third sort of data has potentially immense significance. This comes from giving the civilian research community access to data produced by military satellites. The Global Positioning Satellite system (GPS) gives the possibility of immensely accurate pinpoint localization. Survey satellites carrying banks of different types of visual and other sensors can provide a staggering array of analytic potential.

■ A fourth and final type of contribution to environmental-security research that the armed forces could give immediately lies in management. A good example here is the British Natural Environment Research Council's ambitious AUTOSUB.

Since navies are expert in operating submarines it makes good sense for them to be asked to run the operational side of managing the AUTOSUB fleet for the scientific research community.

Essential to the harnessing of the tremendous potential of existing armed services to modern versions of Captain Cook's voyages is recognition of a principle which the Lord Commissioners of the Admiralty thought it important to state in introducing the account of Cook's voyages.

They made it clear that particular nations undertaking roles of exploration should understand that they were doing it for the general good because,

when plans, calculated to be of general utility are carried into execution with partial views, and upon interested motives, it is natural to attempt to confine, within some narrow circle, the advantages which might have been derived to the world at large by an unreserved disclosure of all that has been effected. And upon this principle, it has too frequently been considered as sound policy, perhaps, in this country as well as among some of our neighbours to affect to draw a veil of secrecy over the result of enterprises to discover and explore unknown quarters of the globe. It is to the honour of the present reign that more liberal views have been now adopted. Our late voyages, from the very extensive objects proposed by them, could not but convey useful information to every European nation; and, indeed, to every nation, however remote, which cultivates commerce, and is acquainted with navigation: and that information has most laudably been afforded. The same enlarged and benevolent spirit, which ordered these several expeditions to be undertaken, has also taken care that the result of their various discoveries should be authentically recorded.

Secrecy has its place and we would do well to heed the views of the eighteenth-century Admirals and recognize that this is not one of them!

Making spin-offs spin

It certainly seems to be the case that, as German and Japanese experiences have demonstrated very fully, the best way to develop advanced technology for civilian use is to develop advanced technology for civilian use. However, the fact is that huge amounts of research and development have been done for military purposes and it is important not to overstate or mis-state the argument about spin-off.

Plainly there are many sorts of militarily-developed or military-stimulated technologies which can have applicability in the fight for environmental security. We have been using energy as our principal theme in this book and so, staying within this fundamental area, let us look at an example of how a technology which would certainly not have reached its present stage of sophistication without the stimulus of military development could be rapidly converted to the promotion of environmental security. Thus this is a sort of parallel to the re-tasking of military forces which we have just described.

The power plant at the heart of the modern warship is the same as the power plant at the heart of a modern warplane. It is the high-efficiency gas-turbine engine. A modern gas turbine is a remarkable example of increasing efficiency. It is a technology which year by year has improved its efficiency since its invention and as the accompanying graph shows it has now reached a very high level both of reliability and of efficiency.

A whole range of advanced technologies has been important in making these improvements in efficiency, not least advanced materials science which has permitted higher and higher operating temperatures within the turbine. The use of carbon fibre was pioneered in aeroplane jet-turbine blades. There is every reason to expect further improvements, because this is one area where the US government has no scruples about investing heavily and directly in research and development. One example is the $3.4 billion, thirteen-year Integrated High Performance Engine Technology programme run jointly by the Department of Defense and NASA.

Gas turbines run on extremely pure fuel. They can also work with a wide range of fuels including gas. If gas is used not in a simple-cycle gas turbine but in one which has steam injection, coupled to an electrical generator, both power and efficiency can be boosted. On a General Electric testbed without steam injection 33 megawatts at 33 per cent efficiency was achieved; with steam injection 51.4 megawatts at 40 per cent efficiency. Even more advanced technology can be used. Using state-of-the-art aero-derivative gas turbines instead of industrial turbines, intercooling between the compressor stages can be employed. Bob Williams and colleagues in the Center for Energy and Environmental Studies at Princeton calculated that, running on natural gas at double the 1986 price, Intercooled Steam Injected Gas Turbines (ISTIG) would produce more power (114 megawatts) and greater efficiency (47 per cent) at costs competitive with a coal-fired steam-electric plant with flue scrubber desulphurization.

But this is just the beginning. In a General Electric Company experiment, a steam-injected gas turbine (STIG) was married to a state-of-the-art coal gasification plant. There are many problems in producing clean gas from coal but they were overcome and a 50 megawatt commercial demonstration was successfully operated.

From the point of view of the move from Stage Three (expensive energy) to Stage Four (renewable energy) described in Chapter 5, the Coal Integrated Gasifier/Steam Injected Gas Turbine (CIG/STIG) plant has the advantage of improving efficiency in the use of a fossil fuel, but the disadvantage that it is still making a net addition to the carbon burden of the atmosphere. What, then, if this technology could be adapted to a renewable source of fuel which had no impact on the biospheric carbon balance?

In pioneering work, Bob Williams and the Princeton group have investigated the technical and economic possibilities of using modern gasification plant and steam-injected gas turbines to run not on coal gas but on *bio-gas*.

Making gas from biomass in a gasifier is in many ways much easier than making clean gas from coal because none of the particulate and sulphur problems associated with coal exist. The bio-gas is particularly clean and therefore it permits state-of-the-art high temperatures in the gas turbine, hence maximum efficiency and power.

The Princeton group have calculated what could be done if these aero-derivative gas turbines, products of the most advanced aspects of the military industrial complex, could be applied under DEFENDUS criteria. They have looked at both a northern hemisphere (temperate) and a southern hemisphere (tropical) example. Their findings are astonishing.

BIG/ISTIG in the North

The European Community wants to take 14 million hectares of marginal land out of cultivation in order to stabilize farm production. If these 14 million hectares were planted under intensive forestry known as SRIC (short rotation intensive cropping) which yields 10 dry tonnes of wood per hectare per year, and if this wood was gasified and burned in high-technology ISTIG-driven district power plants which are typically 50 megawatt capacity and therefore suitable to provide power for a small community, 50 per cent of the EEC's coal-fired power at 1983 levels could be substituted. In the American case, if 40 million hectares of marginal land were planted under SRIC, three-fifths of the United States 1987 coal-fired power could be substituted from this source alone. These are figures to ponder.

When we think back to the possibilities of wind and wave power and of photovoltaic electricity we can begin to see how a full-spectrum sustainable-energy future could be designed. BIG/ISTIG is the technology which can substitute most quickly and elegantly for the dirty lignite-burning power stations of Central Europe. In these newly liberated countries there is little potential for wind and none for wave power. Attempts to build conventional hydro-electric power capacity, most famously between the Czechoslovak and Hungarian shores at the Danube Bend were frustrated, rightly, because the geography was simply wrong. There was little power to get and vast damage to wetland and other sensitive sites to be done. Indeed, resisting the Danube dam in 1989 was one of the catalysts for the democratic revolutions of that year.

However, in Central Europe, there is a great deal of forested land that could be managed sustainably to provide the power supply for BIG/ISTIG. The need to undertake radical improvements in atmospheric pollution is not in question, as we saw in Chapter 4. To promote BIG/ISTIG could become a virtuous circle to replace the pre–

sent vicious circle of environmental damage. At present, the air pollution kills the trees. However, if the trees run the power stations the power stations would not kill the other trees or the people who live in the towns and villages of Central Europe. Nor does the advantage of BIG/ISTIG stop here. For not only does it provide the power required to run society in the present, it is also strong on "inter-generational equity", David Pearce's delightful new phrase which means being considerate to your as yet unborn children's unborn children. This because it produces *zero* CO_2 increase in the atmosphere, which, according to the Princeton calculations for Europe, would lead to a 17 per cent *reduction* in European CO_2 production in the first instance and – because of the increase in the areas under forestation – the sequestering (that is to say locking up in growing trees) of *100 per cent* of the present carbon output of European industry.

BIG/ISTIG in the South

The Brazilian ethanol industry has offered that country the foundation from which it could "leapfrog" the northern industrial world from Stage Three to Stage Four of a stable energy regime. But the enterprise has been put under pressure and indeed under threat by the temporary collapse of oil prices in the mid 1980s. The Princeton group has calculated that BIG/ISTIG could offer the Brazilians tremendous advantages if this technology was integrated with the ethanol production cycle. It works like this.

Traditionally sugar factories have powered themselves by burning the biomass residue from the milling of the sugar cane stalk, called *bagasse*, in small back-pressure steam-turbine co-generation plants which supply typically 20 kilowatt hours of electricity and 400–500 kilogrammes of low-pressure steam per tonne of cane processed. There is interest in using more efficient condensing extraction steam turbines o produce excess electricity. But BIG/ISTIG would make possible a huge leap from 300 to 700 kilowatt hours per metric tonne of sugar cane. Both the *bagasse* and the green top of the sugar plant, called *barbojo*, can be put through a gasifier. If the objective was to maximize the power output from the biomass, in contrast to the usual practice of the sugar factory which is to make the plant *sufficiently inefficient* to burn all the by-produce, thus avoiding the costs of disposal, the plant could generate electricity all year round and could sell its surplus. BIG/ISTIG could produce more than twenty times the current electricity.

In the Brazilian case, it has been calculated that BIG/ISTIG power would be so much cheaper than other sources of electrical power generated in Brazil that it could be sold for general use at a price below that from other sources but generating enough profit to use it to cross-subsidize the cost of ethanol. This would permit the price of ethanol to be reduced to below that of motor fuel, even at the depressed oil prices of the later 1980s. It's another example of a virtuous circle: another free lunch that one is paid to eat.

Steam-injected gas turbines are extremely high technology but none the less are highly appropriate for Third World usage. This is because the engines are robust enough to be fitted in modular form – just as they are in a warship. Nobody tries to mend a gas turbine inside a warship if it breaks down. What happens is that the engine is removed completely, taken to a specialized repair facility and a replacement unit fitted in. Just such a system could be used for power plants scattered throughout the Third World. The result would be *better* availability than from a conventional industrial plant, because of the reduced "down time".

So it's no wild fantasy to see that the

This NASA satellite picture shows huge plumes of smoke rising into the stratosphere from forest burning in Brazil. The extended use of the military's complex surveillance satellite system would provide much greater understanding of the processes that drive our planet. *Right:* this photograph clearly shows the emergence of sand bars in Lake Chad, indicating the steady water loss from the lake.
© NASA.

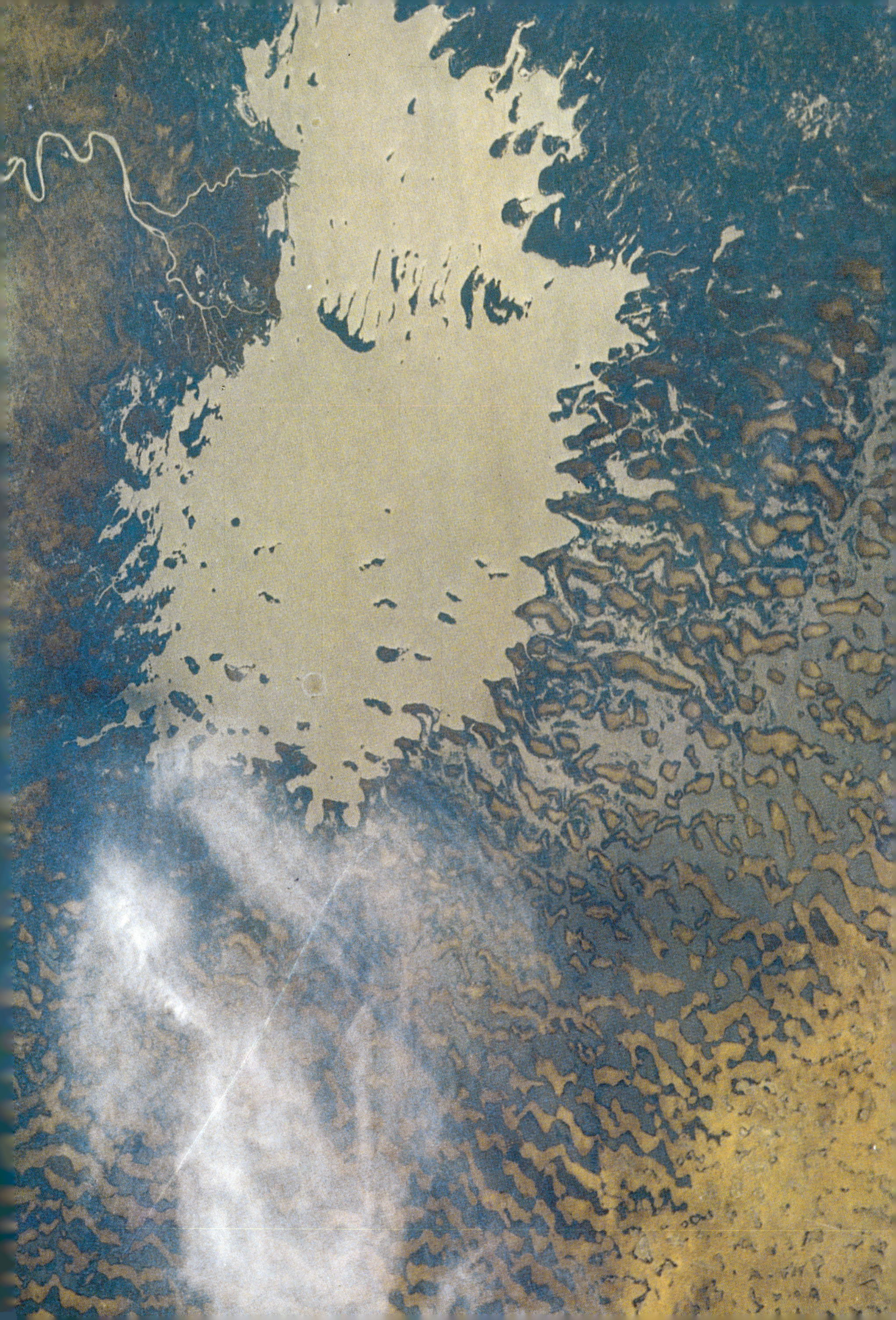

engines which drive fighter planes across the skies of the Middle East can, with the addition of a little extra technology and a great deal of imagination and political will, be used in the front line of the battle for environmental security. Think again of the Top Gun in his Tomcat and the mother with her baby in its pram. Remember how obvious it was that there was nothing that the fighter could do to defend the baby against the unseen environmental threats which assailed it, because there was no enemy at which it could shoot? Now you see that by a process of creative conversion of the skills and technologies that went to make the aeroplane, that isn't true, although it isn't immediately obvious either.

A cautionary tale before we get too excited

In 1983 President Reagan was worried. He was troubled, as many other Americans were, by the fact that the USA's national strategy of defence was based upon a willingness in the last resort to commit acts of indescribable destruction upon the Soviet people at large (with the likely side effect of producing such huge injections of smoke and dust into the atmosphere that the ability of anybody to continue normal life after the nuclear war was in question). This was called "nuclear deterrence", and to Mr Reagan it just didn't seem to square with what he understood to be the underlying values of the American way of life. "The human spirit must be capable of rising above dealing with other nations and human beings by threatening their existence," he mused. Wasn't there another way?

His old friend the nuclear physicist, Edward Teller, inventor of the hydrogen bomb, told him that there was. It was a way which appealed to Mr Reagan because it fitted much better with two of the underlying driving forces of American culture: belief that if you look hard enough you can always find a technological fix, and the desire not to feel ethically uncomfortable in the way that nuclear deterrence made you. What Mr Teller told the President was that the most advanced forms of nuclear technology, rocketry, command, control and communication, expensively joined together, could create weapons that would kill not people but weapons. A great defensive shield in the sky would shoot down Russian missiles before they could fall on American towns and villages and do harm.

The President shared his vision with the American people. "What if free people could live secure in the knowledge that their security did not rest upon the threat of instant US retaliation to deter a Soviet attack; that we could intercept and destroy strategic ballistic missiles before they reached our own soil or that of our allies . . .?"

Ordinary Americans did not disagree. They shared Mr Reagan's values too. So they applauded the next part of his famous "Star Wars" speech. "I call upon the scientific community who gave us nuclear weapons to turn their great talents to the cause of mankind and world peace; to give us the means of rendering these nuclear weapons impotent and obsolete . . . Wouldn't it be better to save lives than to avenge them," he asked rhetorically. The official name of this wizardry was the Strategic Defense Initiative; but soon it was known to its friends and enemies alike as "Star Wars".

From the very outset Star Wars had plenty of enemies. On the morning after the President announced his initiative to the public, Professor Kosta Tsipis of the Massachusetts Institute of Technology was interviewed on a breakfast television show. "Professor Tsipis, you are an expert in nuclear weapons, isn't that right?"

"Yes, I suppose you might say so."

"Professor, tell us what is your view of the President's remarkable new vision?"

"It is forbidden."

"What!" in shocked tone from the interviewer. "How can that be? Who can forbid the President of the United States?"

"The laws of nature," answered the Professor, drily.

He then went on to explain that the plan to use directed-energy weapons to shoot down missiles in space would, in his view, never work because the energized beam would be bent by the earth's electromagnetic field; and nobody had ever made a neutral beam of sufficient power to destroy a missile in flight. The other component of Star Wars, space-borne nuclear-pumped X-ray lasers, had to accomplish a task of accurate aiming and firing in such a short time that even with the facilities of modern computer power the Professor on the morning after the President's speech rather doubted that it could be done. This anyway assumed that the nuclear-pumped X-ray laser would work. If it were to be a chemical laser, which would shoot a beam across space to destroy the rocket, then, said Professor Tsipis dramatically pulling an envelope from his pocket, "I have made a few calculations on the back of an envelope on my way down to the studio. I calculate that to place in space the tonnage of chemical fuel that would be necessary to run the necessary array of space battle stations would require daily flights by the space shuttle for many years into the future."

As time went on another of Professor Tsipis's criticisms on that first morning after the Star Wars speech came to be taken more and more seriously by those in the strategic community, including many who supported the principles of nuclear deterrence. Faced with this horrendously expensive anti-missile system and assuming, against all indications, that it could be made to work to almost perfect efficiency (which was the minimum efficiency permissible given that the destructive power of atomic weapons is such that failure would be allowing any number at all to penetrate the space shield) it would *always be cheaper* for the enemy to overwhelm the expensive shield by simply building more missiles. As Paul Nitze, the Silver Fox of American diplomacy who had written the charter document of the nuclear age (*NSC-68*, see Chapter 1 above) and had been an active player at all stages of Soviet-American affairs from 1953 to 1983, tartly observed, the SDI was not "cost-effective at the margin". In other words it would always be cheaper to make the missiles accelerate faster, to make the missiles mirror plated so that the laser beams bounced off them or simply to build more missiles. The SDI was, in short, a victim of what Herbert York, one of the many distinguished nuclear engineers who in later life became disillusioned with their earlier work, has called the Fallacy of the Last Move: forgetting that there is always a move after the one that you have just made.

So if the SDI did work, its most likely effect at a particularly dangerous time in the Cold War was hugely to stimulate Soviet suspicions, greatly to exacerbate the race in offensive nuclear weapons and thus to produce a net decrease of international security, the very opposite of the hope which Mr Reagan had offered the American people at the outset. If, on the other hand, it did not work, it would none the less represent a gigantic and to many people obscene diversion of precious resources of money and brain power away from any of the many pressing needs on Earth which needed to be addressed.

The Star Wars project was just an extreme example of a general characteristic to be found in the militarized culture of international relations since the end of the Second World War. All expenditures upon arms could be justified on the grounds that they "increased deterrence". In the Star Wars case, the deterrence was supposed to come from the Soviet realization that their weapons would not be able to hurt the Americans, and so they should be deterred by the fact that the Americans could not be coerced.

Increasing deterrence was seen to be a Good Thing. You couldn't have enough of it, because it made your opponent more afraid of attacking (and deterrence-theory always assumes that the opponent wants to attack). The likelihood of war, which in deterrence-theory terms can only arise from an otherwise uncontrollable urge to aggression on the opponent's behalf, was thereby reduced. Other routes to war, like arms races that get out of control, were discounted.

Given the difficulty of making nuclear-deterrence threats credible because of the obvious disparity between any likely real political difference between the two superpowers and the inconceivable destructiveness of any type of nuclear exchange, there was great store placed in the late 1970s and early 1980s upon using the highest of military high-tech to make missiles more accurate and nuclear warheads "smaller" so that destructiveness could be more precisely targeted. The argument was made that nuclear weapons had become usable military weapons again because they could be made "discriminate". Yet as Robert McNamara, who was President Kennedy's Secretary of Defense and the man who presided over the first and biggest upward ratchet of the nuclear arms race in the early 1960s, observed in later life, this fallacy was patent. Conducting arms races in the accuracy of nuclear weapons produced an "action/reaction arms race" which simply took the world a step closer to the brink of Armageddon.

But let's look at the brighter side. The best that any sort of deterrence – nuclear or non-nuclear – can achieve is to stay the hand of a potential aggressor, and at that the historical record is that it hasn't been too successful. Philosophers and nuclear theologians will argue to the crack of doom whether or not the huge nuclear arsenals of the post-war world did or did not have a role in preventing the Soviet Union and the United States fighting each other. Mercifully, this is an issue that is becoming increasingly academic and will no doubt provide long employment for those scholars, but it is of decreasing importance in international affairs. What *is important* is to realize that even if deterrence *does work* it *never* can remove the causes of conflict where they really do exist.

Military power used in a deterrent form is only at the best a palliative. It is something that can be interposed in order to provide a "cooling-off period" until other means can be found to resolve conflict. But because deterrence is based on fear, and because its operation tends to decrease rather than increase understanding of potential enemies, it can never contribute to remove of the cause of war. Nor would any thoughtful military person suggest that it did.

Which is what makes the present moment and the argument made earlier in this chapter so exciting. In Chapter 4 we offered a scenario of how, if we allow denial to freeze our imaginations and fear to drive our minds, we may find ourselves all too easily in a world where environmental stress turns quickly and destructively to war, spitting out refugees.

It has been a theme throughout this book that by all means we must seize now the intellectual and material resources to try and remove the potential causes of such conflict. We must also rebuild the confidence, eroding rapidly and with reason, of worried individuals throughout the industrial world who may not confront the daily horror of living with the Slatkovas in the black triangle of Eastern Europe, but who share a measure of their dread.

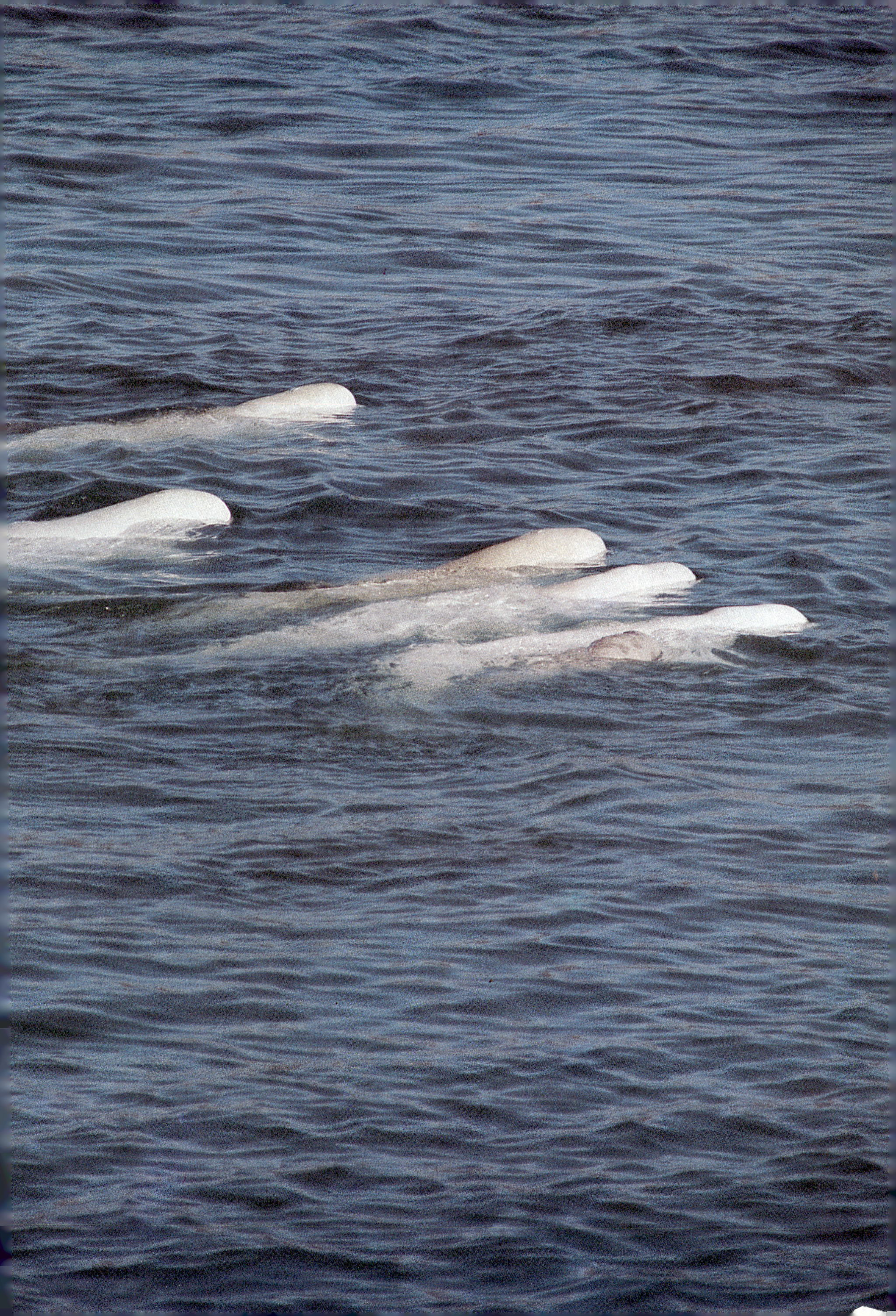

This chapter has explained how, in a paradoxical way, the most exotic fruits of military high technology may be particularly applicable to assisting in that task.

The lesson of Star Wars, as of the earlier stages of the nuclear arms race, was that the more you did it, the worse the situation got. The advanced technology was being fielded under a defective philosophy. That wasn't the fault of the technology as such, but of the context in which it was applied. So it resulted in McNamara's action/reaction downward spiral.

The Top Gun naval fighter pilot can never remove the causes of war, even if he does achieve the maximum destruction in combat. Yet the engine which powers his fighter, taken out of his fighter and put into a power-generation plant driven by bio-gas may help to remove the causes of conflict – oil wars for example – because it can help poor countries to "leapfrog" into a sustainable future and help industrial countries to kick their addiction to huge infusions of fossil-fuel-produced energy, by helping to build robust, ecologically benign sources of power.

President Reagan's first presidency saw the largest increases in American defence expenditure to occur in the history of the republic outside wartime. The Star Wars initiative was at the forefront of that military construction campaign and we can now see that it was quite possibly the last and many think most dangerous example of the old way of thinking about security. To be sure, when spring blossomed in winter in the autumn of 1989 and East Europeans were able to throw off the shackles of repressive Communist regimes, the formal logic of cold warriors and nuclear deterrers had precious little to do with the reasons why that happened.

Rather, as Josef Vavrousek explained in Chapter 4, the extreme environmental degradation of much of Eastern Europe was a powerful stimulant to the anger and eventually to the action of peoples seeking a better life.

Making it safe to cash in the peace dividend

As we said at the beginning of this chapter, we live now in transitional times. As yet there is no evidence that the core military expenditures of the world will be substantially cut – that the "peace dividend" may be drawn on any scale – until the indicators are unequivocally clear that the old types of military threats against which defence was required have diminished substantially. The disarmament treaties of 1989 and 1990 are encouraging pointers to that moment but do not yet represent that moment themselves.

The need for military force will continue to be perceived in the old-fashioned sense so long as potential causes of war persist. Many would argue that the moment will never arise that all military force can be dispensed with.

We argued above that all the means are there to make the United Nations truly the world's policeman, and we believe that this would be a more desirable state of affairs than any single nation, by desire or by force of circumstance, being compelled to take upon itself that role. As Lester Brown of the World Watch Institute said,

> We all have an interest in preserving the habitability of the planet ... this is exciting because it could lead to a far stronger United Nations... whose peace-keeping forces have the capability of intervening quite effectively anywhere in the world. As the founders [of the UN] originally envisaged, it would be possible for governments to relax on the military security front and they could begin to shift resources into environmentally sustainable economic activities.

But we haven't reached that point yet. This doesn't mean that we should not all do the essential thinking as to how we would structure such residual military force under international control. That indeed should be done and it should be one of the benefits drawn from the tragic Gulf crisis of 1990–91 that the international legal and strategic community applies itself with vigour to preparing the United Nations for the activation of the Chapter VII powers which control those activities.

The problem we have addressed in this chapter is how to use existing military forces and their technologies *at this transitional moment* to try and turn the underlying tendencies of the international order away from possible conflict and towards potential collaboration. This battle cannot wait until a peace dividend can be released from the reduction of the world's present arsenals; nor need it. The case for action stands on its own grounds.

In the United Kingdom in 1990 Friends of the Earth UK, jointly with the SaferWorld Foundation, launched what they called the Fifty Percent Initiative. The objective was bold. It was to accomplish a 50 per cent cut in world military expenditures by the end of the millennium. That means an 8 per cent per year reduction in real terms in defence expenditures; interestingly, about the same rate of reduction being considered in the United Kingdom for quite other reasons.

Essential to the Fifty Percent Initiative is the recognition that it has to have two accompanying components. One is a carefully pragmatic strategy for the restructuring of residual military forces so that they would be appropriate to the types of roles that might be required as we moved through times of turbulence ahead. The other is to stress that the resources saved should not be frittered away on general tax cuts or other diffuse or short-term purposes. No, the money saved should be focused upon the environmental-security agenda, thus moving towards a world in which security could be increased in direct proportion to the decrease of expenditures on military preparations of the old-fashioned sort. Spent in the ways we have advocated, that money and those redeployed technologies would erode the old reasons for war as they built an environmentally secure future.

Once the transitional phase has been safely navigated, it becomes not only possible but desirable to diffuse as rapidly as possible throughout society the benefits of ending unproductive labour. Why?

In 1776, in *The Wealth of Nations*, Adam Smith wrote,

> . . . such people as they themselves produce nothing, are all maintained by the produce of other men's labour. When multiplied therefore to an unnecessary number, unproductive hands, who should be maintained by a part only of the spare revenue of the people may consume so great a share of their whole revenue, that all frugality and good conduct of individuals may not be able to compensate the waste and degradation of produce occasioned by this violent and forced encroachment . . . the whole army and navy, are unproductive labourers. Great fleets and armies in time of war acquire nothing which can compensate the expense of maintaining them even while the war lasts.

President Dwight Eisenhower, delivering what became known as his "Cross of Iron" speech in April 1953, spelled out the choices that face us:

> Every gun that is made, every warship launched, every rocket fired, represents, in the final analysis, a theft from those that hunger and are not fed, who are cold and are not clothed."

Conclusion

We have explained that building environmental security is a problem which is at the same time both strange and familiar: strange in that it involves recognizing threats which come from hitherto unexpected quarters – the air, the water, the soil. Familiar in that the means which must be mobilized in order to begin the battle to defend the environment lie close to hand. These means, we have shown, are both *conceptual* and *material*: ironically, many of the ways in which we have analysed threats in the past turn out to be far more appropriate for the analysis of the threats of the future than they were during the period of the Cold War. Many of the technologies developed to build the arsenals of that confrontation between the superpowers turn out to be immediately and creatively and powerfully applicable when drafted into service on the new battle front. But there is more good news.

In the early 1980s, when relations between the Soviet Union and the United States were frozen, many people were filled with dread at the prospect that a nuclear war could come about as a result. They listened with fascinated horror as President Reagan inadvertently joked about bombing the Soviet Union at a press conference, unaware that the microphones in front of him were already switched on.

Faced with this fear, people had the choice of exercising denial or of becoming active. In this book we have seen how powerful, how general, how reasonable and – ultimately – how suicidal the choice of denial has been to many groups of victims in the past. For the vocal minorities who managed to deny denial in the early 1980s, the range of actions which they could undertake was necessarily limited by the nature of the threat which they perceived. What could an individual do to push back the perceived threat of an impending nuclear war? The answer was that they could protest by marching in the streets, protest by writing letters to newspapers, protest by holding public meetings and publishing newsletters, protest by talking about their anxieties to others. All these are activities which have a high "entry price": that is to say, undertaking any of them involves a substantial encroachment upon the way in which one lives one's daily life. So they will not be choices for the majority. Furthermore, there was no "Show me!" crisis. Protesters about nuclear weapons rarely if ever came face to face with nuclear weapons, only with the places where they were hidden away, like Greenham Common. Again, absence of a "Show me!" crisis meant that the majority was likely to be unmoved. Because there is no direct causal link between the symbolic protest action that the individual can undertake and the desired result, it requires exceptional force of character and doggedness to persist in the taking of such symbolic actions against the apparent tide of events. People grew tired and, for whatever reason, nuclear war did not occur.

In the later 1980s, in Eastern Europe, environmental security appeared for the first time as a driving force in "conventional" political change. Josef Vavrousek, now Federal Minister of the Environment in Czechoslovakia:

"The first sort of disturbance was due to the unbearable ecological situation. Horrific air pollution in Prague and other cities. On November 10, 12 and 13 1989 there were large demonstrations about it in northern Bohemia which were broken up by police and soldiers. It was later, on November 17, that events [the Velvet Revolution] took place in Prague. Thus, by the end of 1989, it was obvious that the whole system would not sur-

© NASA

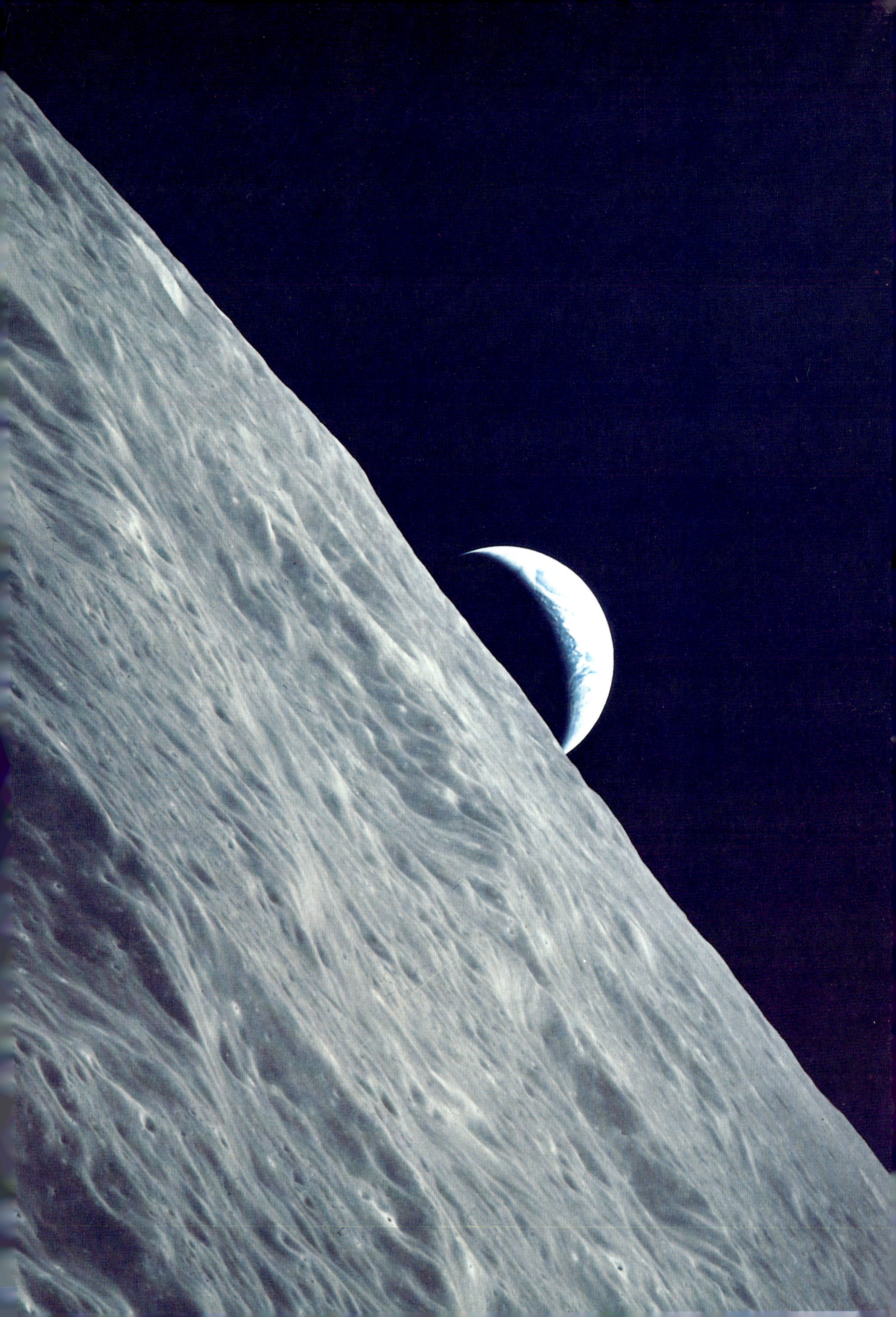

vive long. The only unknown was what would spark off the revolt . . . I believe that had it not been sparked off by the brutal suppression of the student demonstration in Prague of 17 November, it would have been triggered by protests against unbearable deterioration of the environment. We should remember that the winter of 1989 was very mild, as were the two previous ones and that during a mild winter, the problems tend to become worse."

Since the winter of 1989 in Eastern Europe and countries like Czechoslovakia, political circumstances have changed so that there can be open discussion of the depth of environmental pollution. The same freedom of speech gives voice to impatience that these issues be addressed immediately. What we see happening now is that these issues become, for the first time and in the first place in European history, woven into the fabric of civil society.

The choice we all face is the ancient choice which has faced animals and men from time immemorial. Faced with a danger, do you fight or do you flee? Denial advises us to flee. But what does it mean to fight?

The answer is different from that open to people concerned about the danger of nuclear war a decade ago. Our daily actions contribute either positively or negatively to the deterioration of the environment, and because the sources of environmental stress exist at every scale, from the actions of each individual, of each group, each society, each nation and each region upon the planet, real and immediate responses are available at each level.

There is far more that you, the reader, can do about global warming or about depletion of the stratospheric ozone layer than you could do about fears of war between the Soviet Union and the United States leading to nuclear holocaust.

Lester Brown of the World Watch Institute explains:

One of the interesting contrasts between the world order of the last half century, organized around ideological conflict, and the new order, organized around environmental sustainability, is that in the Cold War, the number of people who were directly involved in the military was relatively small . . . in the battle to save the planet, to secure the environmental future of the planet, we are all involved because every life-style decision we make, what we eat, how we travel, how many children we have, all affect the future of the planet. It is very different. We now have to wrestle to change our own values and attitudes and behaviour.

At first it may be a little disconcerting to realize that every tiny decision that we make in daily life is "political" in the politics of the biosphere. Positively or negatively, the decisions that we take by switching off the light or not switching off the light as we go out of the kitchen into the dining room multiplied many millions of times by millions of others, compose our collective if unconsciously expressed view on environmental security.

So the challenge is to make the response conscious. The idea of environmental security helps us to give real substance to the global village. It does so by showing clearly the intricate chains of cause and effect between individual and collective behaviour that affect people who may live very different lives thousands of miles away.

In the battle for environmental security, knowledge is power. It is power to deny denial. It is power to make conscious the unconscious choices which shape our lifestyles. It is power to seize the resources which surround us, designed for a different set of circumstances, and to reorient them, to begin to combat the new threats. It is the power to choose the spring of hope, the age of wisdom, the best of times.

Further Reading

Annuals and publications

Scientific American have produced two special editions: *Managing Planet Earth* (1989) and *Energy for Planet Earth* (1990), which contain articles and excellent graphics by experts in the field. They are very readable and available as books, published in the USA.
(Scientific American, 415 Madison Avenue, New York, NY 10017, USA)

Ruth Leger Sivard produces *World Military and Social Expenditures*, which is currently in its thirteenth edition. It is one of the most accessible guides to comparisons in military and social budgets.
(World Priorities, Box 25410, Washington DC 20007, USA)

UNICEF publish *The State of the World's Children* each year. It is a valuable source of information about a wide range of social and economic indicators.
(Oxford University Press, Walton Street, Oxford OX2 6DP, UK; 200 Madison Avenue, New York, NY 10016, USA)

UNEP – the United Nations Environment Programme – produces a wide range of publications. THE UNEP/GEMS Environment Library is particularly useful.
(UNEP, PO Box 30552, Nairobi, Kenya, with offices worldwide)

The World Resources Institute is another independent policy research centre based in Washington DC. It produces detailed studies and makes policy recommendations on issues ranging from energy policy, debt for nature swaps and reforestation programmes. *Energy for a Sustainable World* by Jose Goldenberg, Thomas B. Johnson, Amulya K. N. Reddy, Robert H. Williams is a key work. Each year WRI produces *World Resources – A guide to the Global Environment*, an authoritative mix of data and analysis.
(WRI, 1709 New York Avenue, NW Washington DC 20006, USA)

The Worldwatch Institute in Washington DC, which is an independent non-profit making research organization, produces an occasional series of papers on a range of environmental issues. Each year they also publish *State of the World, a Report on Progress Toward a Sustainable Society*.
(WW Norton & Co. Inc., 500 Fifth Avenue, New York, NY 1010, USA; WW Norton & Co. Ltd, 37 Great Russell Street, London WC1B 3NU, UK)

Atlases

5000 Days to Save the Planet Edward Goldsmith, Nicholas Hildyard, Peter Bunyard and Patrick McCully
(London: Hamlyn, 1990)

The Gaia Atlas of Planet Management Norman Myers (ed.)
(London: Pan Books, 1985)

The Gaia Peace Atlas – Survival into the Third Millenium General Editor Frank Barnaby.
(London: Pan Books, 1990)

There are very many books on the environment but we recommend three seminal works:
Silent Spring, by Rachel Carson, was the first book to warn people of the poisoning of the environment. First published in America by Houghton Mifflin (1962) and in the UK by Hamish Hamilton (1963), it has been reprinted dozens of times in many languages.

Gaia: A New Look at Life on Earth by James Lovelock. The first book to describe the Gaia hypothesis that the Earth is a living organism in its own right.
(Oxford University Press, 1979)

Our Common Future, the report from the World Commission on Environment and Development – the book that launched the debate about sustainable development.
(Oxford University Press, 1987)

Almost any way you turn there is information to be found about how the individual can contribute to environmental security. "Green" consumer guides on both sides of the Atlantic have sold in large numbers, magazines and newspapers regularly carry features about shopping for a safer world or making your house energy efficient. Television and radio have also been responsive, with a range of programming from children's TV to detailed documentaries focussing on environmental issues.

The Earthscan Action Handbook by Miles Litvinoff contains a wealth of ideas for individual action on environmental issues.
(London: Earthscan, 1990)